Henning Lategahn

Mapping and Localization in Urban Environments Using Cameras

Schriftenreihe
Institut für Mess- und Regelungstechnik,
Karlsruher Institut für Technologie (KIT)

Band 028

Mapping and Localization in Urban Environments Using Cameras

by
Henning Lategahn

Dissertation, Karlsruher Institut für Technologie (KIT)
Fakultät für Maschinenbau
Tag der mündlichen Prüfung: 16. August 2013
Referenten: Prof. Dr.-Ing. C. Stiller, Prof. Dr.-Ing. S. Hinz

Impressum

Karlsruher Institut für Technologie (KIT)
KIT Scientific Publishing
Straße am Forum 2
D-76131 Karlsruhe

KIT Scientific Publishing is a registered trademark of Karlsruhe
Institute of Technology. Reprint using the book cover is not allowed.

www.ksp.kit.edu

Print on Demand 2013

ISSN 1613-4214
ISBN 978-3-7315-0135-0

Mapping and Localization in Urban Environments Using Cameras

Zur Erlangung des akademischen Grades

Doktor der Ingenieurwissenschaften

der Fakultät für Maschinenbau
Karlsruher Institut für Technologie (KIT)

genehmigte

Dissertation

von

DIPL.-INFORM. HENNING LATEGAHN

Tag der mündlichen Prüfung: 16. August 2013
Hauptreferent: Prof. Dr.-Ing. C. Stiller
Korreferent: Prof. Dr.-Ing. S. Hinz

to my wife and son
Eike and Jarno

Acknowledgment

This Phd thesis would not have been possible without so many people who have helped and supported me and at this point I would like to express my gratitude.

Firstly, I would like to thank my supervisor Prof. Dr. Christoph Stiller for the opportunity to complete this work at his institute and the support I have received. I would like to thank my co-supervisor Prof. Dr. Stefan Hinz for the supervision and the fruitful discussions.

The two years I collaborated with the Volkswagen research devision have been a thrilling time and I am happy to have gotten the chance to get to know the department. In particular I want to thank Dr. Jan Effertz, Dr. Thorsten Graf and Dr. Dirk Stüker for their support.

The collaboration with Daimler research during the last two years of my studies have been exciting and seminal. Special thanks go to Eberhard Kaus, Dr. Martin Haueis, Dr. Uwe Franke, Christoph Keller, Dr. Carsten Knöppel and Dr. Jochen Hipp. Beyond that, I want to thank the KIT part of the team for the support and discussions: Julius Ziegler, Markus Schreiber, Philipp Bender and Tobias Strauss. Furthermore, I want thank all colleagues of our institute for the many discussions at the coffee breaks, summer seminars and for the nice spare time activities ranging from playing volleyball to barbecue. Special thanks go to my scrum partners: Dr. Holger Rapp, Philip Lenz and Miriam Schönbein. Moreover, I want to thank Julius Ziegler, Bernd Kitt, Jonas Firl, Dr. Andreas Geiger, Benjamin Ranft, Johannes Beck and Eike Rheder for proof reading an initial version of this manuscript and the excellent feedback. I received a great portion of support from my former student helper Johannes Beck which I would like to acknowledge.

I want to thank our office staff, our system administrator and the workshop staff for their excellent support in all endeavors.

Furthermore, I want to thank my parents in law for continuous encouragement over the past years. Special thanks goes to my parents for having always supported me. Finally, I am deeply grateful to my wife and son whom I dedicate this work to. I want to thank them for countless joyful moments.

Karlsruhe, June 2013   Henning Lategahn

Abstract

Next generation navigation systems and advanced driver assistance system both require a high precision ego localization. A multitude of comfort and safety systems can be realized if such a localization solution is coupled with highly accurate digital maps. In fact, both are prerequisites for automatic driving, a goal that is pursued for decades but still remains unsolved. In this thesis we present a system to fully automatically create a highly accurate visual feature map from image data alone. Moreover, a system for high precision self localization relative to this visual map is presented. Mapping and localization heavily depend on a powerful image processing front end and we present a method to automatically learn a visual descriptor which is tailored to this specific computer vision domain.

The proposed mapping algorithm computes a graph of poses and pairwise motion constraints which is optimized in a least squares sense. We introduce a novel place recognizer which detects areas of self overlap of the mapping surveys. Pairwise image similarities are computed from holistic features and putative loop closure hypothesis are refined to increase robustness to perceptual aliasing by dynamic programming. Thereafter, a large set of 3D landmarks are automatically extracted and stored in an efficient data structure. We demonstrate the feasibility of our approach on several challenging data sets. Four survey trajectories of several ten thousand poses are automatically merged into one consistent pose graph representation.

The localization algorithm which we present requires only a monocular camera and any additional hardware is superfluous. First, the nearest pose of the map is found by an efficient search strategy in the space of appearances. Thereafter, a two step procedure is followed which firstly computes a rough six degrees of freedom ego pose estimate by minimizing the squared back projection error of landmark observations. Secondly, a sliding window of past such estimates is jointly re-optimized to increase robustness and enforce temporal consistency. We demonstrate centimeter-level accurate self localization on several data sets with forward and backward facing cameras. The high precision self localization algorithm allows the implementation of an augmented reality system and we present numerous experimental results of it.

Both of the aforementioned methods largely depend on powerful image descriptors and finding a good descriptor is cumbersome. We address this problem and present a novel method to fully automatically learn an algorithm to visually describe a region around a given pixel position of an image. We identify a set of algorithmic building blocks which allows to create a multitude of descriptors by chaining these blocks in arbitrary order. To the best of our knowledge, we are the first to show how many well established descriptors like local binary patterns (LBP), binary robust independent elementary features (BRIEF), histograms of oriented gradients (HOG) and speeded up robust features (SURF) can be constructed using our blocks. Moreover, we propose a fitness function that evaluates any given descriptor in the realm of place recognition under varying illumination conditions. A genetic algorithm iterates selection and mutation steps and automatically derives a novel descriptor which we dub DIRD (DIRD is an illumination robust descriptor). We assess DIRD on a disjoint test set and demonstrate a significant superiority over its handcrafted counter parts. Thereafter, the point matching performance of DIRD which is of utmost importance for successful localization is evaluated on three test sets exhibiting large illumination variations. DIRD substantially outperforms state-of-the-art alternatives in all tests; by a factor of two in some cases.

Keywords: Localization, Mapping, SLAM, Descriptor Learning, Computer Vision

Kurzfassung

Die nächste Generation der Navigations- und Fahrerassistenzsysteme werden beide eine hoch genaue Eigenlokalisierung benötigen. Eine Vielzahl an Komfort- und Sicherheitssystemen kann realisiert werden, wenn eine solche Lokalisierungslösung mit hoch genauen digitalen Karten kombiniert wird. Beides sind Grundvoraussetzungen für das automatische Fahren; ein Ziel das seit Jahrzehnten verfolgt wird, jedoch bisher ungelöst ist. In dieser Arbeit stellen wir ein System vor, welches rein bildbasiert eine hochgenaue visuelle Merkmalskarte erstellt. Darüber hinaus stellen wir ein Verfahren zur hochgenauen Eigenlokalisierung relativ zur vorherig genannten Karte vor. Sowohl die Kartierung, als auch die Lokalisierung sind stark von einem performanten Bildverarbeitungsfrontend abhängig und wir stellen eine Methode zum automatischen Erlernen visueller Deskriptoren vor, welche für eine spezielle Bildverarbeitungsdomäne maßgeschneidert sind.

Der vorgestellte Kartierungsalgorithmus berechnet einen aus Posen und Bewegungsschätzungen bestehenden Graphen, welcher im Sinne der kleinsten Fehlerquadrate optimiert wird. Wir stellen einen neue Methode zur Ortserkennung vor, welche Selbstüberlappungsbereiche der Kartierungstrajektorie erkennt. Paarweise Bildähnlichkeiten werden von holistischen Merkmalsvektoren berechnet und vorläufige Ringschlusshypothesen werden durch dynamische Programmierung überprüft um die Robustheit gegenüber visuellen Uneindeutigkeiten zu erhöhen. Danach wir eine große Menge an 3D Landmarken extrahiert und in einer effizienten Datenstruktur gespeichert. Wir demonstrieren die Anwendbarkeit unseres Ansatzes auf mehreren anspruchsvollen Datensätzen. Vier Kartierungstrajektorien mit mehreren zehntausend Posen werden automatisch in eine konsistente Posengraphrepräsentation überführt.

Der vorgestellte Lokalisierungsalgorithmus benötigt lediglich eine einzige Monokularkamera und jegliche Zusatzhardware ist überflüssig. Zuerst wird die nächste Pose der Karte durch eine effiziente Suchstrategie im visuellen Merkmalsraum gefunden. Danach wird ein zweischrittiges Verfahren ausgeführt, welches zuerst eine grobe Eigenposenschätzung mit sechs Freiheitsgraden durch Minimierung von Landmarkenrückprojektionsfehlern berechnet. Danach wird eine gefensterte Historie solcher Schätzungen gemeinsam erneut optimiert um Robustheit zu erhöhen und eine zeitliche Konsistenz zu erzwingen. Wir demonstrieren eine zentimetergenaue Eigenlokalisierung auf mehreren Datensätzen mit vorwärts und rückwärts gewandten Kameraaufbauten. Die hohe Genauigkeit der Eigenlokalisierung erlaubt die Umsetzung eines Augmented Reality Systems und wir zeigen mehrere experimentelle Ergebnisse eines solchen Systems.

Beide vorherig genannten Methoden sind stark von performanten Bilddeskriptoren abhängig und es ist schwierig einen guten Deskriptor zu finden. Wir adressieren diese Problem und stellen eine neue Methode zum vollautomatischen Erlernen eines Algorithmus zur visuellen Beschreibung einer Bildregion vor. Wir identifizieren eine Menge von algorithmischen Bausteinen, welche die Konstruktion einer Vielzahl an Deskriptoren durch Verkettung dieser Blöcke erlaubt. Nach bestem Wissen, sind wir die ersten, die zeigen wie etablierte Deskriptoren wie local binary patterns (LBP), binary robust independent elementary features (BRIEF), histograms of oriented gradients (HOG) und speeded up robust features (SURF) durch unsere Blöcke konstruiert werden können. Des Weiteren, stellen wir eine Bewertungsfunktion vor, die einen gegeben Deskriptor auf Funktionalität im Bereich der Ortserkennung unter variierenden Beleuchtungsbedingungen hin bewertet. Ein genetischer Algorithmus iteriert Auswahl- und Mutationsschritte und erlernt so automatisch einen neuen Deskriptor, den wir mit DIRD (DIRD is an illumination robust descriptor) bezeichnen. Wir bewerten DIRD auf einer disjunkten Testmenge und zeigen eine deutliche Überlegenheit über händisch erstellte Algorithmen auf. Danach wird die Performanz von DIRD auf Punktmatchingproblemen, die bei der Lokalisierung von aller höchster Wichtigkeit sind, auf drei Testdatensätzen mit hoher Beleuchtungsvariation, bewertet. DIRD performt deutlich besser als der Stand der Technik auf allen Datensätzen; manchmal um den Faktor zwei besser.

Schlagworte: Lokalisierung, Kartierung, SLAM, Deskriptorlernen, Bildverarbeitung

Contents

Notations and Symbols

Acronyms

ACC	Adaptive Cruise Control
ALOI	Amsterdam Library of Object Images
AR	Augmented Reality
AUC	Area Under the Curve
BRIEF	Binary Robust Independent Elementary Features
DIRD	Dird is an Illumination Robust Descriptor
DOF	Degree of Freedom
EKF	Extended Kalman Filter
EM	Expectation Maximization
GMM	Gaussian Mixture Model
GNSS	Global Navigation Satellite System
GPS	Global Positioning System
HOG	Histogram of Oriented Gradients
LBP	Local Binary Pattern
NLS	Nonlinear Least Squares
PR	Precision Recall
RANSAC	Random Sample Consensus
ROC	Receiver Operating Characteristic
SIFT	Scale Invariant Feature Transform
SIMD	Single Instruction, Multiple Data
SLAM	Simultaneous Localization and Mapping
SURF	Speeded Up Robust Feature

Notations

l_j	Landmark
p_i	Pose

x	State vector
z	Measurement vector
u, v	Pixel position
d	Disparity
$\pi(\cdot)$	Camera projection function
f_i	Holistic feature vector
$\mathbb{I}(\cdot)$	Indicator function

1 Introduction

Modern societies largely depend on mobility for the vast majority of people. Most households in industrialized countries own at least one vehicle and its usage is part of everyones life. This freedom of mobility, however, comes at the price of 4009 fatalities and 392365 injuries caused by traffic accidents in Germany in 2011 [22]. Traffic remains the largest single cause of death in the European Union for ages 15 to 29 [16].

The desiderata for safety is ubiquitous and car manufactures have spent enormous endeavors to increase vehicle safety. In fact, recent developments have led to a pronounced decline in traffic related mortalities. In 1990, 7906 people were killed in traffic in Germany [22] which almost doubles the current number despite an increase in new vehicle registrations [22]. It is largely agreed that this decline is partially attributed to modern active safety systems such as collision warning, adaptive cruise control (ACC), lane departure warning, blind spot detecting and many more. These systems have reached the market and exhibit a high degree of maturity.

Nowadays, much effort is spent on pushing the degree of automation even further. Systems that take over full vehicle control in critical situations seem to be within reach. A partially or fully automatically operating vehicle has many desirable and appealing properties. It cannot suffer from distraction and fatigue, can decide in milliseconds and may possibly cooperate with other traffic participants and thereby increase street capacity without additional infrastructure.

All of the aforementioned systems share a common dependency on both high precision digital maps and a centimeter-level accurate self localization. Static objects of the environment that are persistent over time such as curb stones, traffic signs, lane markings and traffic lights can easily be stored in such maps. During online localization, the relative vehicle position of these objects can be retrieved easily from the ego position at any time. Thereby, the on board environment perception can be moved to an offline computation hence exonerating electronic control units from computationally demanding tasks. Furthermore, the "sensing range" of such static map objects is literally unbound. Moreover, any sensor problems caused occlusion can be elegantly avoided.

Another broad application area is automatic driving which has attracted considerable attention, both in media and research, lately and first working systems have been demonstrated. Maps and a highly accurate ego localization solution

Figure 1.1: An example image which is used to derive a high precision ego pose.

substantially simplify this unsolved problem. A path can be planned using the digital map and the localization is the feedback for the steering and acceleration controller. Moreover, the map can be enhanced by additional information like traffic rules, speed limits, positions of traffic lights and many more.

1.1 Contribution

Herein, we present a system for high precision self-localization using a monocular camera only. The camera is localized relative to a visual landmark map. We show how these maps can be computed from visual information alone and detail the localization algorithm. An example of such a camera image is depicted in Figure 1.1. We use gray scale cameras with a resolution of roughly a half mega pixel and 90° field of view. Finally, we introduce a novel method to automatically construct problem specific image descriptors to learn a powerful localization specific descriptor.

Unlike [59] the visual map is created fully automatically from street level imagery and does not require any additional hardware like [46]. We propose a novel place recognizer that is used to ensure consistency in areas of self overlap. The method is very efficient and much faster to compute than a widely used alternative [17] and is freed from any previous training to create codebooks. Moreover, it is robust to illumination variations caused by different times of day which is an issue that is mostly ignored by recent developments in this area. Preliminary results related to mapping have been published in our works [44, 39, 37].

We present a localization algorithm that requires only a single monocular camera whereas [67] have used bulky and costly laser scanners. We demonstrate centimeter-level precision by an extensive set of experiments on real world data and the achieved accuracy ranges among the most accurate systems to date. The authors of [45] report a localization accuracy of 10cm with laser scanners which we demonstrate to have eclipsed. Beyond that, we obviate the need for a global positioning device for initialization purposes as it is required by e.g. [52]. In contrast to [56], the proposed system estimates a six DOF ego pose which allows the implementation of an augmented reality (AR) system. We have published earlier versions of the method in [38, 42, 43, 41].

Many computer vision problems make use of a visual key point descriptor that describes an image region around a pixel position by a numeric feature vector. Point matching, classification and image retrieval are a few non-exhaustive prominent examples. Recently, a flood of newly developed descriptors have been published most of which are designed with a broad applicability in mind. Finding the descriptor that is best suited for localization and mapping is extremely tedious and cumbersome. That is why, we propose to automatically construct problem specific methods from a set of algorithmic building blocks. We demonstrate how widely used descriptors such as local binary patterns (LBP), binary robust independent elementary features (BRIEF), histograms of oriented gradients (HOG) and speeded up robust features (SURF) can be constructed by the proposed blocks. To the best of our knowledge, we are the first to introduce a unification of existing methods and an easy way to automatically create novel ones. The work of Winder, Brown and co-workers [70, 12] present a set of blocks whose parameters are optimized by Powell's method. The order of blocks however is fixed and only the parameters are optimized. Our methods spans a much greater space and is able to create a richer structure. Unlike [12] we demonstrate that many commonly used descriptors such as LBP, BRIEF, HOG and SURF can be constructed using our blocks. We demonstrate a substantial improvement of an automatically learned descriptor over a set of general purpose ones when applied to place recognition and point matching under challenging lighting conditions. The thus obtained method largely contributes to the overall localization performance. Our previous publications [40, 36] have described some related aspects.

1.2 Thesis Overview

The remainder of this thesis is structured as follows.

Chapter 2 reviews related work. First, common methods addressing the localization problem are presented. Global navigation satellite system (GNSS) approaches suffer from shadowing and multipath propagation and several solutions have been proposed which are briefly mentioned. Map relative localization has recently attracted considerable attention. Laser scanners and cameras are most often used for this purpose. Moreover, our approach to localization is related to simultaneous localization and mapping (SLAM) which has been extensively studied in the past decade. The most influential ideas are shortly re-stated. Lastly, work related to automatically learning a visual image descriptor are reviewed.

Nonlinear least squares (NLS) estimation plays a fundamental role throughout this thesis. **Chapter 3** offers an introduction into the subject. First, the problem is defined as the search for an argument of a vector valued residual function such that it minimizes the squared Mahalanobis norm. Factor graphs are introduced which elegantly provide a representation that is equivalent to the residual function and help to ease the understanding of such problems. A toy example accompanying the entire chapter is successively extended to introduce more advanced topics such as manifolds which are used when state spaces need to be over parameterized. A generally applicable and commonly used trick to reduce a weighted least squares problem to an ordinary one is presented which allows to treat all NLS problems of this thesis as ordinary ones for reasons of better readability whilst silently assuming all noise covariance matrices correctly.

Chapter 4 elucidates the mapping process that is used to compute a visual map entirely from a set of stereo sequences. The proposed method allows to fuse recordings of several survey trajectories into one single pose graph representation. To this end, areas of self overlap are reliably detected by the proposed place recognizer and hence, enforcing consistency in these areas. Correctly detecting areas of self overlap is of utmost importance. It ensures consistency in these areas which would otherwise be unusable for localization purposes. Furthermore, we propose an algorithm that computes a large set of 3D landmarks from the recorded imagery. Salient points are associated between several images and the landmark position is estimated such that it explains the observed pixel positions as good as possible. Finally, the map data structure is presented which is a crucial prerequisite for fast online landmark retrieval. The main concept of this chapter is illustrated in Figure 1.2.

The details of how to localize a single monocular camera relative to the aforementioned visual map are elaborated in **Chapter 5**. The presented algorithm follows a three step approach. First, a very rough pose estimate is determined to initialize all subsequent localization steps. Thereto, a topological localization algorithm is proposed which finds the pose of the map that is closest to the current ego position. This nearest neighbor search is entirely performed in the space of appearances as no global positioning device is used. After topological localization all 3D landmarks of the immediate vicinity of the current camera pose are loaded from disk and associated with salient points of the current camera image. These associations are harnessed to derive a rough metric ego pose estimate which is referred to as one-shot estimate. Finally, a set of past one-shot estimates are re-optimized jointly

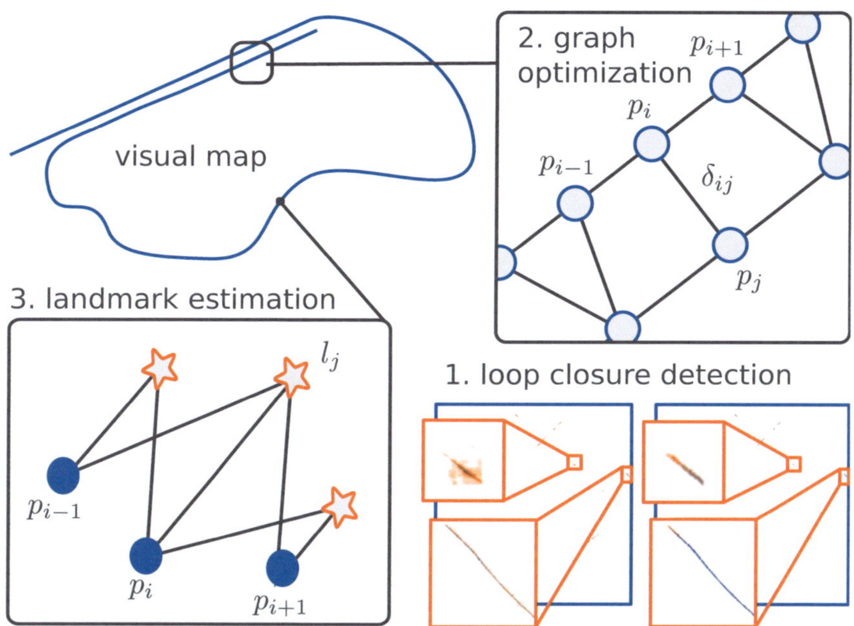

Figure 1.2: The mapping process creates the visual map used for localization and is introduced in Chapter 4. First, a visual odometry induced pose graph is constructed. Areas of self overlap are reliably detected by the proposed loop closure detector which matches overlapping subsequences by dynamic programming. Lastly, natural landmarks are detected and their spatial position is estimated by NLS estimation using the result of the aforementioned pose graph optimization.

to increase overall robustness and this third processing step is dubbed pose adjustment. An overview of the chapter is depicted in Figure 1.3

A framework for automatic learning of descriptors is presented in **Chapter 6**. A set of algorithmic building blocks is proposed which can be chained in any order to form a single descriptor for each such combination. This construction scheme allows to automatically assemble a multitude of different descriptors. A fitness function that evaluates the performance of a descriptor in the realm of place recognition under varying illumination conditions is proposed. Finally, the construction scheme is coupled with the fitness function to derive a meta heuristic that fully automatically finds a descriptor that performs particularly well in the domain of this specific problem. We demonstrate a substantial improvement of the learned descriptor over state-of-the-art general purpose ones. The key concept is shown in Figure 1.4.

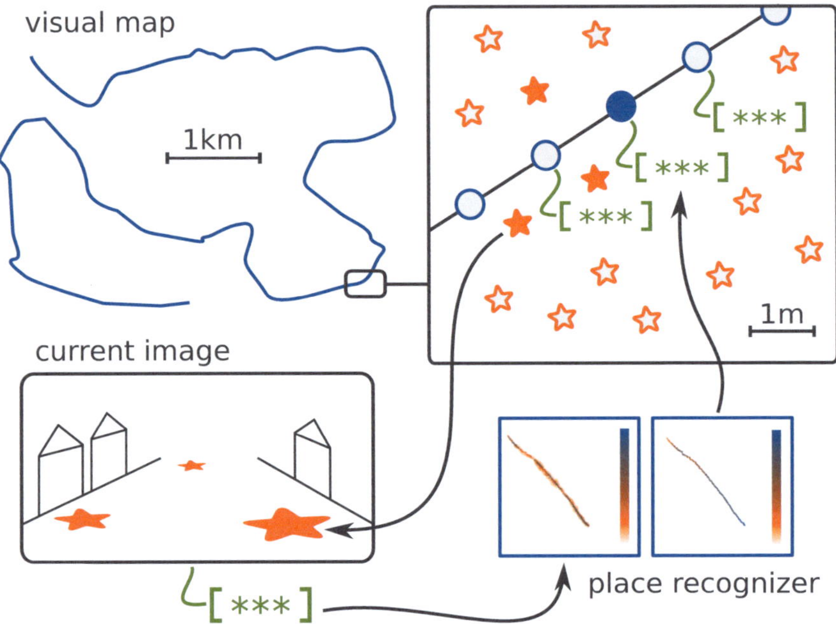

Figure 1.3: Our novel localization method is elaborated in Chapter 5. First, the current camera image is translated into a space of appearances and the pose of the map that is most similar is found using a place recognizer. Thereafter, landmarks in the vicinity of the vehicle are loaded from disk and associated with pixel positions of the current camera image. Landmark associations are finally harnessed to derive a high precision six dimensional ego pose.

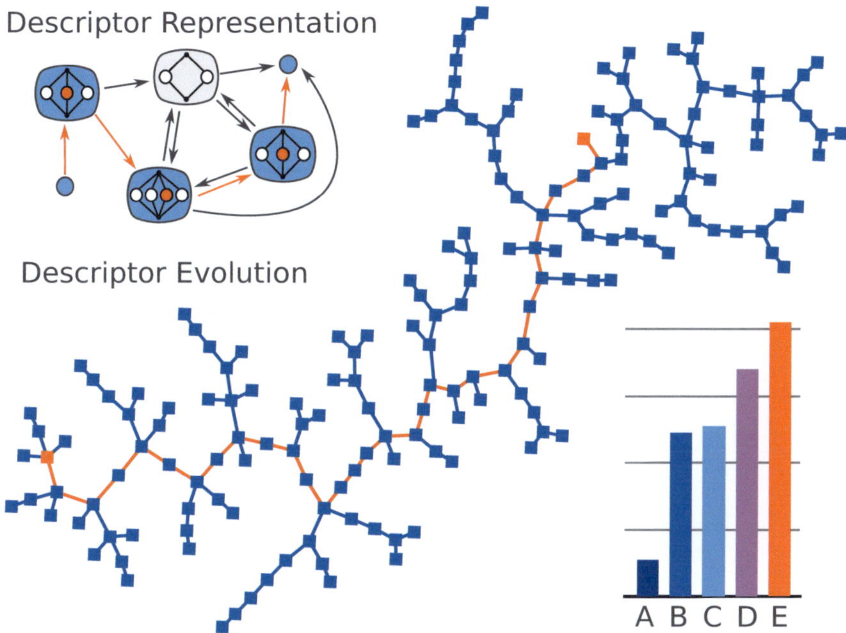

Figure 1.4: We introduce a set of algorithmic building blocks which allow to assemble descriptors by chaining these blocks in arbitrary order as described in Chapter 6. If these blocks are nodes of a graph then a single descriptor is fully described by a path trough this graph (top left). An evolution strategy is used to iteratively mutate a root descriptor to yield successively better ones. One such evolution is shown (middle). The thus learned descriptor (E) outperforms state-of-the-art alternatives (A-D) in place recognition under varying illumination conditions as shown on the right bottom.

2 Related Work

In this chapter we will review related work. First, we cover state-of-the-art localization methods in Section 2.1. Thereafter, we present a summary of the most influential simultaneous localization and mapping (SLAM) methods and systems in Section 2.2. Finally, work related to descriptor learning is elucidated in Section 2.3.

2.1 Localization

In this section we review recent advances related to localization in general and map relative localization in particular. We first briefly touch GNSS approaches and how these can be improved by integrating road maps into the estimator. Thereafter, we delve into the realm of map relative localization in 2D and in 3D. Finally, we shortly sketch topological camera based localization methods.

A straight forward and commonly followed approach to address the localization problem is to use GNSS. Pseudorange measurements are fused in a filter framework to estimate the global position and clock offset of the receiver. Multipath propagation, however, causes severe problems and can cause catastrophic divergences of the estimator. To mitigate these effects Sünderhauf and co-workers developed a robust least squares solver which jointly estimates the receivers position and the set of valid pseudorange measurements in [65]. To this end, one switch variable is estimated for every pseudorange measurement jointly with the position state vector. These switch variables reveal any outlying measurement. A greatly improved accuracy is demonstrated using a low cost global positioning system (GPS) receiver albeit not reaching the accuracy necessary for some applications.

Using road maps in conjunction with GNSS seems an appealing possibility to constrain the ego position. Several approaches have recently emerged and we exemplarily cite the work of Drevelle and Bonnifait [21]. The search for the most likely ego position is recast as a constraint satisfaction problem. Interval analysis is used to find a set of solutions (ego positions) that satisfies GPS measurements and map induced restrictions. Several hypothesis can be computed in cases of ambiguities. The reported localization accuracy is 6.5 meters in 95% of the cases in a street canyon like scenario in downtown Paris.

Integrating additional knowledge (the map in this case) contributes to the overall accuracy of the localization system as demonstrated by the work mentioned above. Li et al. [47] have followed a similar idea and have added additional cues into their localization estimator. GPS readings, maps and beyond that, a monocular camera is used. The camera detects lane markings and road signs of a known position. All three complementary measurements are finally fused by a particle filter to yield a refined ego pose.

Recently, precomputed maps have been used for localization purposes. The obtained localization is always relative to a previously recorded map and global position accuracy therefore depends on the map accuracy. However, map relative accuracy is often the desired goal for e.g. path planning and the like. Furthermore, any error prone handling of occluded or reflected satellite reception is obviated completely. A laser range finder is used in [67] to localize within a 3D polygon map. A particle filter propagates the posterior distribution over the ego pose through time.

A rotating laser scanner is used for localization in [45, 46]. Levinson and collaborators have proposed to use infrared remittance values of laser beams of the road surface only. The road surface can be found rather accurately in laser point clouds and is likely to be persistent over time. SLAM approaches are used to smooth the map and enforce consistency in areas of self-overlap whereas particle filters are their choice of localization estimator.

Laser scanners of this type are prohibitively expensive on the one hand and cause severe packaging problems on the other. Hence, their use in series production vehicles is inadmissible. The recent explosive growth of imaging technologies offers a solution and cameras are used by Napier and Newman in [51]. A model of the road surface is computed by computing local orthographic projections of it from stereo data. This road surface model is stored as a map and subsequent passes of the same route are localized relative to it. The mutual information between synthesized views of the surface and the current camera image is maximized to derive a 2D ego pose.

Pink [56] follows a related idea. Aerial images are pre-processed by detecting corners of road markings. Unlike [51], point features (landmarks) are used in his approach. An online camera image is searched for these landmarks and iterative closest point algorithms yields the estimated 2D ego pose. Finally, these ego position estimates are fused with a motion model to constrain the motion and increase overall accuracy.

The aforementioned systems localize in 2D. Some application areas, however, require a 3D localization. The notion of landmarks for a 6 degree of freedom (DOF) localization is also presented in [69]. A micro aerial vehicle is localized relative to a dense point cloud map. A sparse point cloud map is computed by structure from motion first and gaps are filled in by patch based multi view stereo yielding the dense map. Virtual camera images can then be computed from the dense representation. The ego pose can thereafter be recovered from point correspondences with the current online image.

A very recent 6 DOF localization system using a monocular camera is presented in [2] by Alcantarilla, Dellaert and others. In fact, their system shows some resemblance to the work presented herein albeit some pronounced differences. A visual map is computed by means of bundle adjustment using stereo cameras. Landmarks are extracted and their visual description is stored for localization. An NLS solver is used to yield the ego pose estimate during online operation from a monocular camera. The initialization process of finding a proximity of the map to localize in differs from ours and scales much worse. In fact, their system is designed for small indoor scenarios whereas our system scales to large scale outdoor trajectories. Moreover, the carefully designed visibility prediction process of [2] is completely obviated by our map data structure.

Cameras have also been used by Badino in [3, 4]. Imagery was recorded for an urban area and holistic image features describing each single pose of the mapping trajectory are extracted from the images and stored as the map. During online localization current image features are matched to the map and position estimates are smoothed by fusing odometry information. The method is dubbed topometric localization. The final position estimate will always correspond to exactly one pose of the mapping trajectory.

The aforementioned methods only seem to be the tip of the iceberg in the realm of holistically describing images for this kind of localization. Milford [50] has pushed the idea further by describing panorama images by only a few bits. Place matches are computed for double round trip trajectories of lengths up to 70km. The dynamic time warping of the pairwise image difference matrix appears to be a crucial ingredient. Our place recognizer draws some inspirations from this work.

2.2 SLAM

SLAM is the long known problem of localizing a robot within a map while computing that map at the same time. Localizing is enabled by known maps while map generation depends on a localization solution. Hence, SLAM tries to solve this elusive chicken and egg problem. Most often maps are represented by a collection of landmarks which are sensed by some sensor. Bayesian filters like extended Kalman filters (EKFs) aim at estimating a state comprised of all landmarks and the current ego position [5, 23, 19]. Sensor readings can be fully predicted from one such state vector and compared to the actual sensor output to yield the filter innovation.

Despite its theoretical soundness, the filter approaches as stated above suffer from well known Draconian limitations in scalability preventing large scale real time systems. Discovering a special structure of the state covariance matrix which exhibits strong correlations only between landmarks that have been sensed jointly eventually led to submapping approaches. Only a small fraction of the state vector and covariance is updated at each time step and a global update is postponed as long as possible (e.g. until a loop closure occurs). These methods have constant complexity most of the time. We exemplarily mention the work of Pinies and co-workers [55]. Their solution is numerically equivalent to the regular EKF formulation after every global update and does not necessitate any approximations.

A long known solution from photogrammetry experienced a resurgence of interest once computing power had increased: bundle adjustment. All robot poses and all landmark positions are stacked into one joint state vector which is estimated by nonlinear least squares (NLS) estimation. Levenberg-Marquardt and Gauss-Newton methods are popular choices of solvers. The measurement matrix of the linearized system of equations which is iteratively solved exhibits an extremely sparse structure. The emergence of sparse matrix solvers using variable reordering [1] finally led to the breakthrough. A relative representation of the problem dubbed relative SLAM is presented in [62]. Triggs et al. offer a good introduction into bundle adjustment in [68].

Nowadays the landmark/pose notion of SLAM is replaced by a simple pose-only surrogate. The state of the map consists of all poses of the robot trajectory and the motion induced pose graph is estimated from sensor readings. Once loop closures are introduced, the system of pose to pose constraints becomes overdetermined. This pose graph is then an abstract representation and agnostic of the sensor that created it. It is finally solved by standard nonlinear least squares machinery. The removal of landmark positions from the problem allows to estimate very large trajectories [20]. A fine introduction into the subject is presented in [28]. A flexible open source software library is presented in [35]. Most approaches of solving pose graphs can be traced back to the influential work of Lu and Milios [49].

Another problem tightly coupled to SLAM is the detection of previously visited places dubbed loop closure detection. FabMap presented in [17] by Cummins and colleagues has emerged into the work horse of loop closure detection. It applies an appearance based approach. A probabilistic model of places is learned from salient image features. Large feature vocabularies need to be trained beforehand. Their work is robust to perceptual aliasing albeit being computationally rather expensive. The feature extraction is quite time consuming. To mitigate the effects of visually describing a multitude of image features in every image Sünderhauf and co-workers [66] have resorted to a simplistic approach. The image is down sampled and partitioned it into small equally sized image tiles each of which is holistically described by only one single image descriptor. Concatenating single tile features into one yields the descriptor representing the appearance of the entire image. Place recognition is thereafter straight forwardly achieved by nearest neighbor search in the space of appearances.

2.3 Descriptor Learning

We present a descriptor learning framework and review related literature here. The work closest to our descriptor learning framework is the work of Brown and co-workers [70, 12]. A set of blocks are presented which are combined to assemble a descriptor. The block parameters are optimized by Powell's method. Blocks include smoothing, non-linear transforms, pooling and normalization. A large training set of different views of the same points is created by large scale bundle adjustment. The order of blocks, however, is fixed and only the parameters are optimized. Our methods spans a much greater space of descriptors with more processing blocks. Furthermore, the order of blocks is optimized as well.

Parameters of scale invariant feature transform (SIFT) and histograms of oriented gradients (HOG) features are optimized by the method of [64]. A set of patches are automatically extracted from street level imagery similarly to [12]. The parameter space of these hand crafted methods is thereafter searched for an optimum with car classification in mind. Experiments show a substantial improvement of the trained parameter set over the default set. Furthermore, it is shown how classification accuracy can benefit from application specific parameters.

Philibin et al. [54] and Carneiro [14] both present a method for feature learning. Their work focuses on learning a distance function for image retrieval which is referred to as distance metric learning. Linear projections and deep believe networks are used. We refrained from learning a mere distance transform but rather optimized the entire descriptor.

Descriptors used for image retrieval largely depend on spatial pooling and vector coding. Boureau and colleague [11] systematically investigated the effects of proper pooling and coding choices. By pairwise combination of these choices, a rich set of novel descriptors can be created. The importance of appropriate choices for these steps could be highlighted. They find that sparse coding is more appropriate than soft and hard quantization and maximum pooling substantially improves over average pooling.

Using a filter operation followed by estimating the filter distribution has been investigated in our previous work on texture description in [40]. We test several filter banks (gradient filters, Haar wavelets) and model the filter response by Gaussian mixture models (GMMs) using the expectation maximization (EM) algorithm. Thereby we achieve state of the art texture classification performance.

A large body of literature has been published on evaluating different image detectors and descriptors. It is far beyond the scope of this thesis to exhaustively review them all. We exemplary mention the recent work of Gauglitz and co-workers [25] as this kind of evaluation studies are only loosely related to descriptor learning. A vast test set of images for feature tracking was created and commonly used descriptors are evaluated in terms of matching performance. However no alteration or automatic feature construction is proposed.

3 Nonlinear Least Squares Estimation

Nonlinear least squares (NLS) problems play a fundamental role throughout the rest of the thesis. In this chapter we review the basics of NLS estimation.

First, we give a formal definition of NLS problems and show how these problems can be elegantly modeled by factor graphs. The factor graph notation will be frequently used in the localization Chapter 5 and the mapping Chapter 4. It concisely summarizes the structure inherent to the specific problem and augments the associated error sums in a complementary way. Thereafter, we review numeric algorithms commonly used to derive an estimate to these problems. Then a method is presented that can handle estimates of non-Euclidean spaces such as e.g. rotations in 3D. A manifold approach is reviewed which has gained popularity in the computer vision and robotics community lately. Since NLS problems are notoriously susceptible to any outliers in the measurement vector, we shortly elaborate common solutions to this problem. Finally, we bridge the gap between NLS estimation and Extended Kalman Filters (EKF) which are interpreted as a sequential approximation of the full NLS estimate. The chapter is accompanied by a simple toy example which is successively extended to cover the aforementioned aspects.

3.1 Definition

NLS problems frequently arise in situations where a hidden state vector $x \in \mathbb{R}^n$ needs to be computed from noisy measurements $z \in \mathbb{R}^m$. One example application is model or curve fitting. If a physical phenomenon can be described by an equation containing parameters and a set of noisy measurements is available then these parameters can be estimated. In a usual case the number of measurements outnumbers the number of parameters by a fair margin. Since no parameter set can then be found to fully explain all measurements (with vanishing residuals) a set of parameters is sought that minimizes a squared error sum.

Formally let

$$r : \mathbb{R}^n \to \mathbb{R}^m \tag{3.1}$$

$$x \mapsto \begin{pmatrix} r_1(x) \\ \vdots \\ r_N(x) \end{pmatrix} \tag{3.2}$$

be a vector valued function with $r_i(x) \in \mathbb{R}^{m_i}$ being subvectors of $r(x)$ such that $\sum_{i=1}^{N} m_i = m$. We refer to $r(x)$ as residual vector and x denotes the state. An NLS problem is now the search for an estimate of x that minimizes the squared norm of the residual vector. The estimate is denoted by

$$\hat{x} = \arg\min_{x} \left\{ r(x)^T r(x) \right\} \tag{3.3}$$

$$= \arg\min_{x} \left\{ ||r(x)||_2^2 \right\} \tag{3.4}$$

$$= \arg\min_{x} \left\{ \underbrace{\sum_{i=1}^{N} ||r_i(x)||_2^2}_{=:E(x)} \right\} \tag{3.5}$$

and $E(x)$ is a scalar error function whose minimum is sought. In the model fitting example above the residual vector may be defined as $r(x) = (h(x) - z)$ with $h(\cdot)$ being a measurement function which predicts a measurement from any given state vector x. It is often possible to compute a measurement from a state but not vice versa. Hence one needs to solve for $\hat{x}$ to revert some measurement z, a problem that can be solved by NLS estimation.

We give a simple example next. Suppose there exists a subway train that drives from one subway station A to the next B. A and B are connected by straight tracks and the distance between them is known exactly. The train starts to travel and a device on board measures the traveling distance over time intervals of one minute. The total traveling time shall be N minutes in this example. Hence we obtain N distance measurements $d_1, \ldots, d_N$ for the N intervals and the train arrives at B after the last measurements. The device, however, suffers from some inaccuracies and the measured distances are noisy. The goal shall now be to estimate the positions of the train for every full minute after all measurements have been acquired. A position $p_i \in \mathbb{R}$ is defined as a scalar value of the totally traveled distance until then. The first and last position is constrained to correspond to A and B respectively that is $p_0 = 0$ and $p_N = D$ with D being the known distance between A and

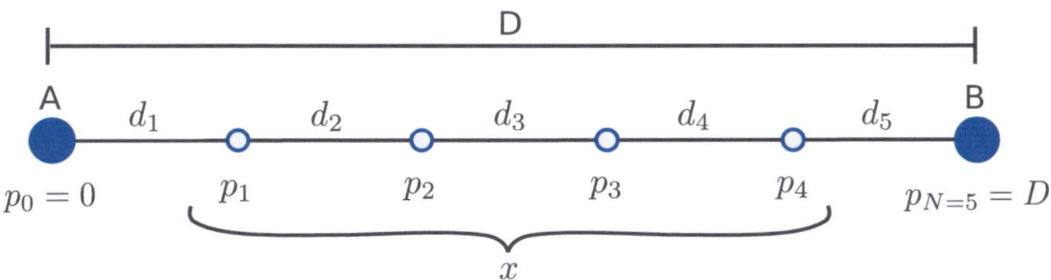

Figure 3.1: The positions of the subway train are schematically depicted. The train starts at the known position A and travels to the known position B. A device measures the traveled distance d_i between any two consecutive positions p_{i-1} and p_i. The NLS problem seeks the positions $p_1, \ldots, p_{N-1}$ such that the noisy measurements d_i are best explained.

B. An example instance of this problem is shown in Figure 3.1. We wish to derive a good estimate of the positions in between. Thus, the state vector corresponds to the positions

$$x = \begin{pmatrix} p_1 \\ \vdots \\ p_{N-1} \end{pmatrix} \in \mathbb{R}^{N-1} \tag{3.6}$$

and is $N - 1$ dimensional. The residual vector is now

$$r(x) = \begin{pmatrix} (p_1 - p_0) - d_1 \\ \vdots \\ (p_N - p_{N-1}) - d_N \end{pmatrix} \in \mathbb{R}^N \tag{3.7}$$

and penalizes any deviation of the distance between two consecutive poses from the measured distance. It yields the error function

$$E(x) = \sum_{i=1}^{N} || \underbrace{(p_i - p_{i-1}) - d_i}_{r_i(x)} ||_2^2 \tag{3.8}$$

whose minimizing argument $\hat{x}$ corresponds to the position estimates of the subway train between the stations A and B.

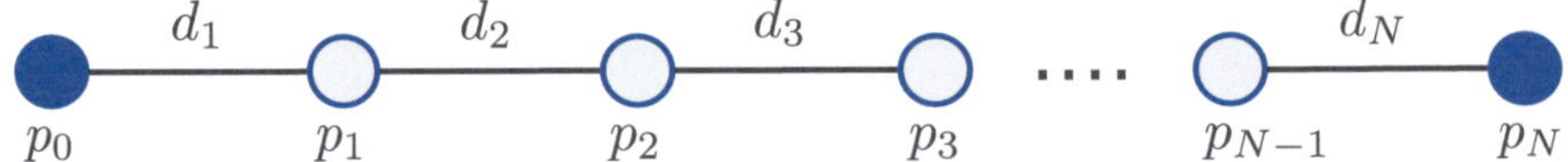

Figure 3.2: The factor graph of the initial toy example is shown. It consists of all variables that appear in the NLS problem. Hollow variables are optimized (comprise the state) and solid ones are fixed and remain unaltered.

3.2 Factor Graphs

Before delving into the technicalities of how to actually compute $\hat{x}$ we introduce factor graphs [34]. A factor graph is a graph that consists of vertices and edges that connect exactly two vertices. Vertices correspond to variables that appear in the error function $E(x)$. A subset of the nodes comprise the state vector. Edges that connect vertices correspond to residuals $r_i(x)$. Figure 3.2 shows the associated factor graph of the train example. One edge exists for every $r_i(x) = (p_i - p_{i-1}) - d_i$ and the summation of (3.8) extends exactly over all edges of the graph. Edges are often labeled with measurements as is in this example. The edge between p_{i-1} and p_i is labeled with the associated measured distance d_i. Finally, all vertices that appear hollow in the graphs are subject to optimization and comprise exactly the state vector x. Solid vertices are variables that cannot be changed during optimization. In this example this holds for the first and last positions p_0 and p_N which remain unaltered. Note the great similarity of the factor graph of Figure 3.2 and the overview sketch of Figure 3.1.

Every NLS problem can be translated into its associated factor graph and vice versa. Therefore, we always present both the error sum $E(x)$ and the factor graph for better readability. In the above example, one edge connects exactly two vertices. In fact, this is the case for all NLS problems throughout the thesis. In general, an edge can connect more than two vertices. However, this never happens in any of the NLS problems of this thesis and we have therefore presented this slightly simplified notation. An extension to graphs that allow to connect more than two vertices is straight forward and is usually referred to as a hyper-graph. A typical example which requires a hyper-graph representation is the joint estimation of scene structure, camera poses and a coordinate transform from measured pixel observations and some pose priors. If the pose priors are given in a different coordinate system (e.g. vehicle coordinates as opposed to camera coordinates) then the transform from vehicle to camera coordinates can be estimated jointly with the scene structure. However, to predict a single pixel observation the corresponding pose, point and coordinate transform needs to be known rendering it a hyper edge (connecting three variables).

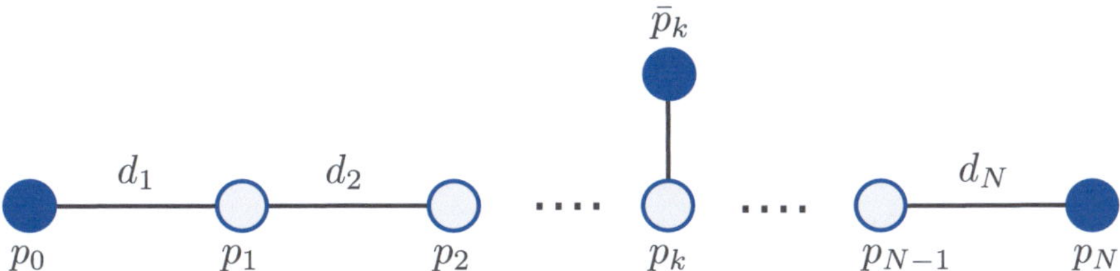

Figure 3.3: The factor graph of Figure 3.2 is slightly extended and an additional prior term is added. This prior is found as an additional summand in (3.9).

We slightly extend the aforementioned example to show how prior information can be integrated into NLS problems seamlessly and how this is shown in the factor graph representation. We suppose that at one point the train passes a landmark of known position and this is detected by the on board device. Hence it is known that some specific pose p_k is nearby the landmark of known position $\bar{p}_k$. To integrate this information the error function (3.8) is extended to

$$E(x) = ||p_k - \bar{p}_k||_2^2 + \sum_{i=1}^{N} ||(p_i - p_{i-1}) - d_i||_2^2 \tag{3.9}$$

which additionally penalizes the deviation of p_k from its prior $\bar{p}_k$. The extended factor graph is shown in Figure 3.3. The prior $\bar{p}_k$ is added to the graph. It is connected by an additional edge with p_k. This additional edge represents the additional summand of (3.9). The prior $\bar{p}_k$ is shown by a solid vertex as it cannot be changed during optimization. This toy example obviously serves only to show the general principle of how prior information can be integrated into NLS problems.

3.3   Normalization

At this point we want to draw the attention to another important issue which has been completely neglected so far. Suppose the subway train device mentioned above provides a quality measure for each of its measurements. That is, that some measurements are known to be more accurate than others. This information must not be ignored during optimization. Let λ_i be the certainty of the distance measure d_i. For increasing λ_i more confidence should be put on d_i. This integrates into the error function as follows

$$E(x) = \sum_{i=1}^{N} \lambda_i ||(p_i - p_{i-1}) - d_i||_2^2 \tag{3.10}$$

$$= r(x)^T \Lambda r(x) \tag{3.11}$$

with $r(x)$ of (3.7) and a diagonal weight matrix $\Lambda = \mathrm{diag}(\lambda_1, \ldots, \lambda_N)$. In general Λ does not need to be diagonal but is required to be symmetric and positive definite. In fact, the error distribution of $r(x)$ is assumed to be Gaussion with covariance matrix Σ leading to the weight matrix $\Lambda = \Sigma^{-1}$ or $\lambda_n = \frac{1}{\sigma_n^2}$ for a diagonal $\Sigma = \mathrm{diag}(\sigma_1^2, \ldots, \sigma_N^2)$. Then the Cholesky decomposition $\Lambda = L^T L$ can be computed and an auxiliary residual function

$$r'(x) = Lr(x) \tag{3.12}$$

can be defined. Finding x that minimizes the squared norm of $r'(x)$ then corresponds to

$$\hat{x} = \arg\min_{x} \{E'(x)\} \tag{3.13}$$

$$= \arg\min_{x} \{r'(x)^T r'(x)\} \tag{3.14}$$

$$= \arg\min_{x} \{r(x)^T L^T Lr(x)\} \tag{3.15}$$

$$= \arg\min_{x} \{r(x)^T \Lambda r(x)\} \tag{3.16}$$

finding the solution of the weighted version as desired. Henceforth, we always assume non-weighted (ordinary) NLS problems for the sake of simplicity but silently apply the aforementioned normalization procedure where necessary.

The possibility to integrate weights allows to finely adjust the influence of prior believe in any estimate (cf. (3.9)).

3.4 Solvers

In this section we present common approaches to solving problems like (3.5). A good introduction into a broad spectrum of optimization algorithms and others is presented in [58]. We derive the Gauss-Newton and Levenberg-Marquardt algorithms next. Both share a common overall structure. At first an initial guess x_0 of the state vector is used for initialization. Then a small change Δ is found such that $x_0 + \Delta$ yields an improvement of the residual norm which is $E(x_0 + \Delta) < E(x_0)$. The process is iterated by setting $x_{i+1} = x_i + \Delta$ until a termination criterion is met. Common choices of termination criteria are to monitor the change in residual norm $E(x_i) - E(x_{i+1})$ or to check the norm of the update vector $||\Delta||$ to decide for termination. Usually the iteration is also terminated once a maximum number of iterations is reached which may hint at an ill conditioned problem.

It remains to show how to compute the update vector $\Delta \in \mathbb{R}^n$ with n being the dimension of the state space. To this end we define an auxiliary function

$$s(\delta) = r(x_i + \delta) \tag{3.17}$$

for a fixed x_i. Note that $s(\cdot)$ behaves like $r(\cdot)$ and can be well approximated by Tailor series expansion for small δ by

$$s(\delta) \approx s(0) + S(0)\delta \tag{3.18}$$

with the Jacobian

$$S(0) = \left. \frac{\partial s(\delta)}{\partial \delta} \right|_{\delta=0} \tag{3.19}$$

$$= \left. \frac{\partial r(x)}{\partial x} \right|_{x=x_i}. \tag{3.20}$$

An update Δ is found by minimizing

$$s(\delta)^T s(\delta) \approx (s(0) + S(0)\delta)^T (s(0) + S(0)\delta) \tag{3.21}$$

$$= s(0)^T s(0) + 2s(0)^T S(0)\delta + \delta^T S(0)^T S(0)\delta \tag{3.22}$$

for δ by equating the derivative of (3.22) with zero. This yields

$$0 = 2S(0)^T S(0)\delta + 2S(0)^T s(0) \tag{3.23}$$

with a unique solution

$$\delta = -\left(S(0)^T S(0)\right)^{-1} S(0)^T s(0) \tag{3.24}$$

$$=: \Delta \tag{3.25}$$

which we take for Δ. Note the difference in notation between δ and Δ. Equation (3.22) is a scalar function with the dependent variable vector δ which has a unique minimum at $\delta = \Delta$ which is a concrete value. The introduction of the auxiliary function $s(\cdot)$ may seem overly complicated and unnecessary at this point but will help to remain consistent once the state vector is of a non-Euclidean space and element of a manifold. The full Gauss-Newton algorithm is shown in Algorithm 1. Another very popular choice is to extend the Gauss-Newton method into

---

Algorithm 1 Gauss Newton

Require: initial guess x_0
Ensure: x that is a local minimum of $r(x)^T r(x)$
  $x \leftarrow x_0$
  while termination criterion not met **do**
    define function $s(\delta) := r(x + \delta)$
    $S \leftarrow \left.\frac{\partial s(\delta)}{\partial \delta}\right|_{\delta=0}$
    solve $S^T S \Delta = -S^T s(0)$ for Δ
    $x \leftarrow x + \Delta$
  end while

---

the Levenberg-Marquardt algorithm shown next. The main difference is the intro-

---

Algorithm 2 Levenberg-Marquardt

Require: initial guess x_0
Require: initial value of $\lambda > 0$
Ensure: x that is a local minimum of $r(x)^T r(x)$
  $x \leftarrow x_0$
  while termination criterion not met **do**
    define function $s(\delta) := r(x + \delta)$
    $S \leftarrow \left.\frac{\partial s(\delta)}{\partial \delta}\right|_{\delta=0}$
    solve $\left(S^T S + \lambda \operatorname{diag}(S^T S)\right) \Delta = -S^T s(0)$ for Δ
    $x' \leftarrow x + \Delta$
    if $r(x')^T r(x') < r(x)^T r(x)$ **then**
      $x \leftarrow x'$
      decrease λ
    else
      increase λ
    end if
  end while

---

duction of a damping factor λ when solving the linearized system of equations. It helps to escape some local minima (but not all) and is generally more robust than Gauss-Newton. However, its convergence speed is a little bit slower in general. Sometimes the Levenberg-Marquardt algorithm is referred to as a trust region method.

3.5 Manifolds

Many times it is necessary to estimate a vector of a non-Euclidean space. Prominent examples are rotations in 3D. Representing rotations by Euler angles causes sever problems in some situations. If the difference of two such vectors needs to be computed then the simple Euclidean norm is deceptive. The vector norm of the vector difference of e.g. $x_1 = (\pi, \pi, \pi)$ and $x_2 = (-\pi, -\pi, -\pi)$ is non-vanishing though they obviously represent the same rotation. Moreover, there exists configurations where one degree of freedom is lost. Suppose the angles represent the rotation around the x-axis, the then rotated y-axis and finally around the twice rotated z-axis. Then, one degree of freedom is lost once the second rotation angle approaches $\frac{\pi}{2}$ as the first and last axis perfectly align and the sum of their angles is the last remaining degree of freedom. This may cause the estimator to diverge once it approaches these configurations during optimization. This problem is referred to as gimbal lock.

A common approach is to use a parameterization that does not suffer from gimbal lock. Unfortunately, there exists no minimal (3-vector) representation of the group of rotations in 3D that satisfies this requirement. Hence, one needs to resort to a suitable parameterization using rotation matrices or quaternions. Thoughtlessly using this parameterization in the state vector is doomed to fail. The algorithms presented above do not assure that the updated state vector $x_{i+1} = x_i + \Delta$ actually represents a rotation matrix or quaternion which both only exhibit three degrees of freedom each. In this section a method that elegantly circumvents this deficiency which has gained popularity in robotics and computer vision is presented. It uses the notion of manifolds which we introduce next.

A manifold is a union of open subsets of $\mathbb{R}^n$ for some n with associated functions for each subset that map elements of the subset to another space $\mathbb{R}^m$. The function that maps between the subsets $\mathbb{R}^n$ to $\mathbb{R}^m$ are called charts. An nice introduction example may be the manifold of a sphere surface. Consider the position of points on the earth surface (assumed to be a perfect sphere). Naturally they are expressed by a 3-vector in some Euclidean coordinate system of three axis. However, only points that belong to the surface are contained in this manifold (hence a union of subsets). For a local region we can simply chop out the region and draw a 2D map of it (say a map of Europe). Within this map we can consistently work with 2D coordinates. The chart then maps from 2D map coordinates to 3D coordinates.

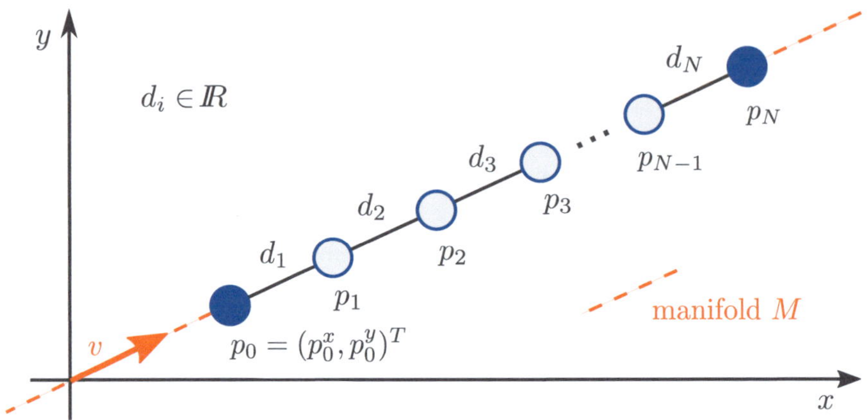

Figure 3.4: The poitions of the toy example are now in 2D having an x and y coordinate. However, all positions are required to lie on the track defined by the line of direction v. The set of all points of this line is the manifold M.

In particular we can compute the difference of two 3D points that belong to the surface and express the difference by 2D coordinates. Moreover, a 3D point can be updated by a small 2D update vector by using the chart (map).

We resume the subway example but now consider each position of the train to be estimated to be 2-vector which is constrained to lie on a line (namely the tracks) of known parameterization. Figure 3.4 shows a sketch. The line manifold is the set of all points of the line

$$M = \left\{ (x, y)^T = \lambda v \,|\, \lambda \in \mathbb{R} \text{ and } v = (v^x, v^y)^T \right\} \tag{3.26}$$

with known direction vector v. The new state vector is now comprised of positions which are an element of this line manifold. Each position is represented by two spatial coordinates $p_i = (p_i^x, p_i^y)^T \in M$ thus

$$x = \begin{pmatrix} p_1^x \\ p_1^y \\ \vdots \\ p_{N-1}^x \\ p_{N-1}^y \end{pmatrix}. \tag{3.27}$$

The chart ϕ of the manifold maps each element to its minimal one dimensional representation by

$$\phi(p_i) = \lambda_i \tag{3.28}$$
$$= \frac{p_i^x}{v^x} \tag{3.29}$$

with $p_i \in M$ and $\lambda_i \in \mathbb{R}$. The inverse finally maps from the minimal representation to the manifold by

$$\phi^{-1}(\lambda_i) = \lambda_i v \tag{3.30}$$

and we can switch back and forth between both parameterizations. Much like in the earth surface example above we can define the difference of two poses of the manifold by the $\ominus$ operator which will yield the difference in minimal representation by

$$p_j \ominus p_i = \phi(p_j - p_i) \tag{3.31}$$
$$= \frac{p_j^x - p_i^x}{v^x} \tag{3.32}$$
$$= \delta. \tag{3.33}$$

Conversely, a small update $\delta \in \mathbb{R}$ can be applied to any pose p_i of the manifold by the $\oplus$ operator by

$$p_i \oplus \delta = p_i + \phi^{-1}(\delta) \tag{3.34}$$
$$= p_i + \delta v \tag{3.35}$$
$$= p_j \tag{3.36}$$

and it follows that

$$p_j \oplus (p_j \ominus p_i) = p_j \tag{3.37}$$

holds. Using the notation of $\oplus$ and $\ominus$ we can restate the residual function (3.7) for the manifold case yielding

$$r(x) = \begin{pmatrix} (p_1 \ominus p_0) - d_1 \\ \vdots \\ (p_N \ominus p_{N-1}) - d_N \end{pmatrix} \in \mathbb{R}^N \tag{3.38}$$

where the regular minus $-$ is replaced by $\ominus$. Finally, we are able to define the auxiliary function for a given state x by

$$s(\delta) = r(x \oplus \delta) \tag{3.39}$$

$$= \begin{pmatrix} (p_1 \oplus \delta_1) \ominus p_0 - d_1 \\ (p_2 \oplus \delta_2) \ominus (p_1 \oplus \delta_1) - d_1 \\ \vdots \\ P_N \ominus (p_{N-1} \oplus \delta_{N-1}) - d_n \end{pmatrix} \tag{3.40}$$

$$= \begin{pmatrix} (p_1^x + \delta_1 v^x - p_0^x)/v^x - d_1 \\ ((p_2^x + \delta_2 v^x) - (p_1^x + \delta_1 v^x))/v^x - d_2 \\ \vdots \\ (p_N^x - (p_{N-1}^x + \delta_{N-1} v^x))/v^x - d_N \end{pmatrix} \tag{3.41}$$

with $\delta = (\delta_1, \dots, \delta_{N-1})^T$ and $\delta_i \in \mathbb{R}$ being a minimal representation. The point is that $s(\cdot)$ is now a vector function of the minimal representation whereas $r(\cdot)$ is a function of the manifold. The Gauss-Newton method is now presented in algorithm 3 that also works for the manifold case. The only difference is the

---

Algorithm 3 Gauss Newton on manifolds

Require: initial guess $x_0 \in M$
Ensure: $x \in M$ that is a local minimum of $r(x)^T r(x)$
    $x \leftarrow x_0$
    while termination criterion not met **do**
        define function $s(\delta) := r(x \oplus \delta)$
        $S \leftarrow \left. \frac{\partial s(\delta)}{\partial \delta} \right|_{\delta=0}$
        solve $S^T S \Delta = -S^T s(0)$ for Δ
        $x \leftarrow x \oplus \Delta$
    end while

---

definition of the auxiliary function $s(\delta) = r(x + \delta)$ versus $s(\delta) = r(x \oplus \delta)$ and the update equation $x \leftarrow x + \Delta$ versus $x \leftarrow x \oplus \Delta$. By construction it is guaranteed that $x \leftarrow x \oplus \Delta \in M$ is still a member of the manifold.

For a state vector that represents e.g. a rotation matrix the gimbal lock problem is fully avoided as long as the state update δ which is still represented by Euler angles is small (and hence far from any lock) which is an assumption that is well fulfilled in practice. Computing the difference between the initial examples of $x_1 = R(\pi, \pi, \pi)$ and $x_2 = R(-\pi, -\pi, -\pi)$ where $R(\cdot, \cdot, \cdot)$ is a rotation matrix is then $x_1 \ominus x_2 = \phi(x_2^{-1} x_1) = (0,0,0)^T \in \mathbb{R}^3$ with $\phi(\cdot)$ being a chart that maps from rotation matrices to Euler angles. Thus, the desired result is obtained by using the manifold of rotation matrices.

3.6 Robustness

NLS problems suffer from a high susceptibility to any implausible measurements often referred to as outliers. These measurements exhibit a much greater noise level than expected. If these measurements remain undetected the solvers of Section 3.4 diverge catastrophically. Many times one single outlier leads to an overall poor estimate. Hence, it is of paramount importance to reliably detect these outliers. Once these measurements are identified, they are removed from the residual function and the outlier cleaned version can be solved as explained above.

A straight forward approach is to investigate the residual function. If certain elements of the residual function that correspond to one measurement are unnaturally high it strongly hints at an outlier. Formally let $r_i(x)$ be a subvector of the residual function $r(x)$ that involves the measurement z_i which is to be inspected. If Λ_i is the inverse covariance matrix of r_i then the squared Mahalonbis norm $R = r_i(x)^T \Lambda_i r_i(x)$ should follow a χ^2 distribution. If $1 - \int_0^R \chi^2(x)dx \ll 1$ holds then it is likely that z_i is an outlier. This method is usually referred to as naïve outlier rejection.

The approach is quite fast to compute in practice and shows acceptable performance if only few outliers are to be expected in the measurements. In situations where the outlier density increases it may not be possible to reliably detect all outliers correctly. A widely used alternative to the problem is the use of a random sample consensus (RANSAC) [24]. A novel randomized version is presented in [15]. First, a minimal set of measurements that is required to compute an estimate is randomly drawn and the state is estimated. Thereafter, all measurements are inspected whether they support the current state hypothesis. The process is repeated and the largest inlier set is used for a final joint optimization. RANSAC requires considerably more time to derive an estimate (due to to its repeated estimation) but achieves excellent results in practice.

RANSAC, however, exhibits one deficiency. For some NLS problems a sensible minimal set of measurements cannot be found. Pose graph estimation as it is presented in Chapter 4 is such an example. Recent advances try to integrate the classification of measurements into inlier and outlier sets into the NLS estimation. Sünderhauf et al. [65] have proposed to augment the NLS problems by one switch variable for every measurement that is possibly unreliable. The switch variable is passed through a logistic function and indicates an inlier by a value near one and outlier by a value near zero. Hence, the residual function can yield a lower minimum by automatically disabling certain measurements. The convergence speed of these augmented NLS problems seem to be a little worse (due to much more complex fitness surface) but provide a powerful alternative to RANSAC.

3.7 Extended Kalman Filter

A common approach that is often followed when estimating a hidden state that evolves over time is the Extended Kalman Filter (EKF). Let $x_0, \ldots, x_N$ be the evolution of the hidden state for consecutive time steps. A transition function $f(\cdot)$ that models the state development is assumed to be known such that $x_i = f(x_{i-1}) + \epsilon_i$ with an unknown noise vector ϵ_i. The transition function is used to predict a future state and may require a control parameter which we neglect here for simplicity. Moreover, there exists a believe in the initial state which is denoted by $\bar{x}_0$. Beyond that, a measurement is obtained for every time step such that $z_i = h(x_i) + \gamma_i$ with known measurement function $h(\cdot)$ but unknown noise γ_i. The goal of the EKF is to recursively estimate the state x_i from the previous state estimate $\hat{x}_{i-1}$ using the state prediction $f(\hat{x}_{i-1})$ and the current measurement z_i. The state x could, for example, encode the state of a moving car-like vehicle and may contain position, yaw, angular and linear velocities. As such vehicle state cannot change arbitrarily a future state can be predicted for small time lags by extrapolating the vehicle movements given the aforementioned parameters. Any change in velocities is then modeled as (system) noise ϵ. A measurement device for such a system could be a GNSS receiver which outputs a noisy position reading z which obviously depends on the state. An EKF filter can then be used to infer the state (velocities etc.) from the GNSS readings for any time step. The graphical model is depicted in Figure 3.5. Note that the graphical model is not equivalent to the associated factor graph. The graphical model is generic and holds for all EKF problems whereas a factor graph can only be constructed for one specific NLS problem.

The goal of EKF estimation is to recursively obtain an estimate $\hat{x}_i$ of the current state x_i. To this end, the previous state estimate $\hat{x}_{i-1}$ is used to predict the current state by $f(\hat{x}_{i-1})$. Thereafter the state prediction can be refined by exploiting the current measurement z_i to yield the final state estimate $\hat{x}_i$. System (ϵ_i) and measurement noise (γ_i) are assumed to be independent, zero mean and Gaussian distributed with covariance matrices Q_i and R_i respectively. In EKF filtering, it is crucial to track the certainty in the current state estimate, too as it influences the amount of change during an update step from measurements. The estimated state covariance of $\hat{x}_i$ is denoted by P_i.

Next we present both the prediction and update step for the EKF. On initialization the state is set to $\hat{x}_0 = \bar{x}_0$ and P_0 is set to $\bar{P}_0$ which is provided with $\bar{x}_0$. The

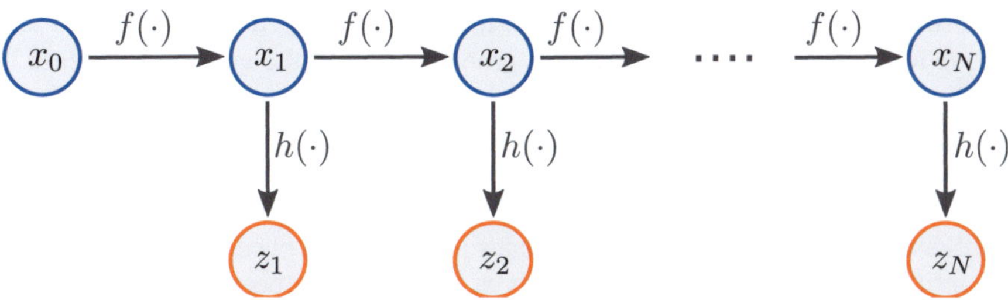

Figure 3.5: The graphical model of an EKF is shown. An initial state x_0 is propagated over time by the transition function $f(\cdot)$. The state, however, is only implicitly observed using the measurement function $h(\cdot)$ yielding the measurements z_i.

prediction of the state covariance is

$$P_i^p = F_{i-1}P_{i-1}F_{i-1}^T + Q_i \tag{3.42}$$

where F_{i-1} is the Jacobian of $f(\cdot)$ for $\hat{x}_{i-1}$ and Q_i is the system noise covariance as above. The state is simply predicted by applying $f(\cdot)$ to the current state estimate

$$\hat{x}_i^p = f(\hat{x}_{i-1}). \tag{3.43}$$

The update step thereafter fuses the current state measurement z_i with the state prediction $\hat{x}_i^p$ and the predicted covariance P_i^p by

$$y_i = z_i - h(\hat{x}_i^p) \tag{3.44}$$
$$S_i = H_i P_i^p H_i^T + R_i \tag{3.45}$$
$$K_i = P_i^p H_i^T S_i^{-1} \tag{3.46}$$
$$\hat{x}_i = \hat{x}_i^p - K_i y_i \tag{3.47}$$
$$P_i = (I - K_i H_i)P_i^p \tag{3.48}$$

where I is the identity matrix and R_i is the measurement covariance matrix.
The EKF summarizes all past measurements in one single state vector $\hat{x}_i$ and recursively updates that state from noisy state measurements z_i. In the case of a constant dimension of the state vector (thus when the state vector does not grow over time) the update requires constant time for each update step independent of the number of measurements. This very appealing complexity property largely

contributes to the success of the EKF which has developed into a standard technique for a broad range of estimation problems. As in the general NLS case the EKF assumes outlier free measurements. Outlier detection can be handled by e.g. gating techniques which reject very unlikely state measurements.

Next, we show how the same problem can be solved using NLS estimation. To this end, we define the state of the NLS problem to be $x = (x_0^T, \ldots, x_N^T)^T$. Hence, the NLS estimate aims at estimating all states jointly and at once [61]. Of course this is only possible once all measurements $z_1, \ldots, z_N$ have been acquired. Though, a sequential version to estimate all states until the current time i can be derived straight forwardly. The residual function is now given by

$$r(x) = \begin{pmatrix} x_0 - \bar{x}_0 \\ f(x_0) - x_1 \\ \vdots \\ f(x_{N-1}) - x_N \\ h(x_1) - z_1 \\ \vdots \\ h(x_N) - z_N \end{pmatrix} \tag{3.49}$$

whose squared norm is minimized by the algorithms of Section 3.4. Just like the EKF the estimated state shall not deviate too much from its prediction ($f(x_{i-1}) - x_i$), the initial state shall be near its believe ($x_0 - \bar{x}_0$) and the deviation of the predicted measurements from the actual ones ($h(x_i) - z_i$) is penalized. Note that the initial state covariance of $\bar{x}_0$, the system and measurement noise can be handled by the normalization method of Section 3.3.

Moreover, we note that a single EKF update is in fact very similar to solving an NLS problem for the residual function

$$r_{\text{ekf}}(x_i) = \begin{pmatrix} f(\hat{x}_{i-1}) - x_i \\ h(x_i) - z_i \end{pmatrix} \tag{3.50}$$

which contains the current EKF state as the NLS state x_i [8]. Here $\hat{x}_{i-1}$ is the EKF-estimated previous state and $f(\hat{x}_{i-1})$ is the prediction of the current state. The EKF, however, must properly propagate the state covariance as well.

Finally we note that it is possible to define the EKF state to contain a fixed number of past states $x_{i-K}, \ldots, x_i$ in the state vector jointly. The EKF then becomes very similar to NLS estimation except that the EKF does not iterate during the update step as Gauss-Newton and Levenberg-Marquardt algorithms do. In fact, it is even possible to iterate the update step which is yet another flavor and referred to as the iterated EKF (IEKF). Using this variant allows to finely adjust the parameter K

(the time lag) to balance accuracy and runtime. The IEKF solution approaches the full NLS solution for $K \to i$.

4 Mapping

We present our mapping algorithm next. The aim of the mapping process is to obtain a map which allows to localize a camera relative to it. Our goal is to create such a map from stereo vision alone. Since large areas may need to be covered, the algorithm is designed to handle discontinuous recordings. In particular a map can be constructed from surveys of different days. Overlap between single or multiple trajectories are robustly identified and the map is extended accordingly. Hence, a map can be extended every time a new dataset is added to the database. Furthermore, the mapping procedure is fully automatic and requires no manual intervention. The obtained map has no global reference since only vision is used to compute it. However, some cases may require a geographic reference of the entire map. Therefore, we propose a slight extension to add such geo reference to it. This step, however, is optional and is only mentioned for completeness. A GPS receiver is required only in that case.

Throughout the rest of this chapter we assume a set of survey trajectories with associated stereo image sequences to be available from which we derive the visual map. At first, these trajectories are computed by visual odometry and made consistent in areas of overlap by means of loop closure detection and pose graph optimization. Thereafter, landmarks are extracted from the associated stereo image streams and their spatial position is estimated. Finally, the entire map data needs to be stored efficiently on secondary memory since its size easily eclipses available primary memory capacities. The aforementioned steps are presented detailedly in the following sections.

A preliminary work of the place recognizer has been published in [36]. A similar landmark estimator has been presented in [42, 41].

4.1 Pose Graph Estimation

We assume a set of survey recordings $\mathcal{S}_1, \ldots, \mathcal{S}_R$ each of which contains a sequence of stereo images $\mathcal{S}_r = \left(s_1^r, \ldots, s_{N_r}^r \right)$. These sequences cover the area for which a map is sought and may be from different days. To spare the reader a flood of indices we make a notational simplification and assume only one recording which consists of the concatenation of all stereo images

$$\mathcal{S} = \left(s_1^1, \ldots, s_{N_1}^1, s_1^2, \ldots, s_{N_R}^R\right) =: (s_1, \ldots, s_N) \tag{4.1}$$

which is still discontinuous at the transitions from one sequence to the next (e.g. $s_{N_1}^1$ to s_1^2). We do not lose generality by renaming the elements of the sequence (4.1). It still contains the stereo images of multiple sequences and days.

Before delving into the details of the mapping tool chain we need to slide in a short paragraph on notations. Throughout the rest of the thesis we assume poses to be parameterized by 4×4 homogeneous matrices

$$p = \begin{pmatrix} \mathbf{R} & \mathbf{t} \\ \mathbf{0}_{1\times 3} & 1 \end{pmatrix} \in SE(3) \tag{4.2}$$

with 3×3 rotation matrix $\mathbf{R}$ and 3×1 translation vector $\mathbf{t}$. Moreover, the chart $\phi(\cdot)$ maps from this over parameterized manifold $SE(3)$ into $\mathbb{R}^6$ the minimal parameterization of 3D angle and 3D translation vector [31]. The motion operator $\oplus$ which applies a motion $\delta \in \mathbb{R}^6$ to a pose $p \in SE(3)$ is defined by

$$p_2 = p_1 \oplus \delta \tag{4.3}$$
$$= p_1 \cdot \phi^{-1}(\delta) \in SE(3). \tag{4.4}$$

Conversely, the subtraction of two poses yields a change by

$$\delta = p_2 \ominus p_1 \tag{4.5}$$
$$= \phi(p_1^{-1} \cdot p_2) \in \mathbb{R}^6. \tag{4.6}$$

A good introduction into this subject can be found in [31] and [63].

First, we show how to compute a pose graph $\mathcal{G}$ from the sequence (4.1) of stereo images. A pose graph is a graph that consists of a set of poses each of which is associated with exactly one stereo image of one recording. These poses are interconnected by motion constraints. Formally, a pose graph for the N stereo images of $\mathcal{S}$ is defined as the two tuple

$$\mathcal{G} = \left(\underbrace{\{p_1, \ldots, p_N | p_n \in SE(3)\}}_{\text{set of poses}}, \underbrace{\{\delta_{ij} \in \mathbb{R}^6 | i, j \in \{1, \ldots, N\}\}}_{\text{set of pairwise constraints}} \right) \tag{4.7}$$

with δ_{ij} being a motion estimate between pose p_i and p_j.

The graph is constructed by adding the pose to pose motion constraints δ_{ij} first. If pose p_i is in close proximity to pose p_j the stereo images s_i and s_j can be used to estimate the motion δ_{ij} by means of visual odometry [26]. Poses p_i and p_j are assumed to be spatially close to each other if they are successive poses of the same initial sequence (i.e. $j = i+1$). For $j \neq i+1$ these poses are assumed to be close if the loop closure detection of Section 4.2 has identified the stereo images s_i and s_j to show the same place. We postpone the technicalities of this place recognition until the next section.

The final step of pose graph estimation is to compute the poses of the graph such that it matches all pairwise constraints as good as possible. One may expect $p_j = p_i \oplus \delta_{ij}$ or equivalently $\delta_{ij} = p_j \ominus p_i$ which cannot be satisfied for all δ_{ij} due to the motion δ_{ij} being a noisy estimate. Hence, each motion constraint induces an error $e_{ij} = p_j \ominus p_i - \delta_{ij}$ which is non-vanishing. Thus, the pose graph optimization seeks a configuration of all poses such that the sum of squared norms of all these errors is minimal. The following error function

$$E(p_1,\ldots,p_N) = \sum_{ij} ||e_{ij}||^2 \tag{4.8}$$

$$= \sum_{ij} ||p_j \ominus p_i - \delta_{ij}||^2 \tag{4.9}$$

reflects this error where the summation extends over all i,j for which one δ_{ij} exists. The result of this NLS problem

$$\hat{p}_1,\ldots,\hat{p}_N = \underset{p_1,\ldots,p_N}{\arg\min} \{E(p_1,\ldots,p_N)\} \tag{4.10}$$

is taken as the poses of the visual map. These estimates are kept fixed after estimation and used throughout the rest of the mapping process. They are also stored as part of the map data structure (see Section 4.4).

Note that (4.9) exhibits a degree of freedom which needs to be handled. An arbitrary rotation or translation of all $\hat{p}_1,\ldots,\hat{p}_N$ yields the exact same error value (4.9). We follow a common approach and simply fix exactly one pose to an arbitrary value during optimization. A judicious choice is to set p_1 to the coordinate origin.

Figure 4.1 shows the factor graph of (4.9). A simple sample trajectory is shown which contains all the poses. Poses are interconnected by motion constraints. The first pose is depicted solid which means it is kept fixed during optimization. All other poses are hollow and comprise the state of (4.9). The summation of (4.9) again extends over all edges of the factor graph.

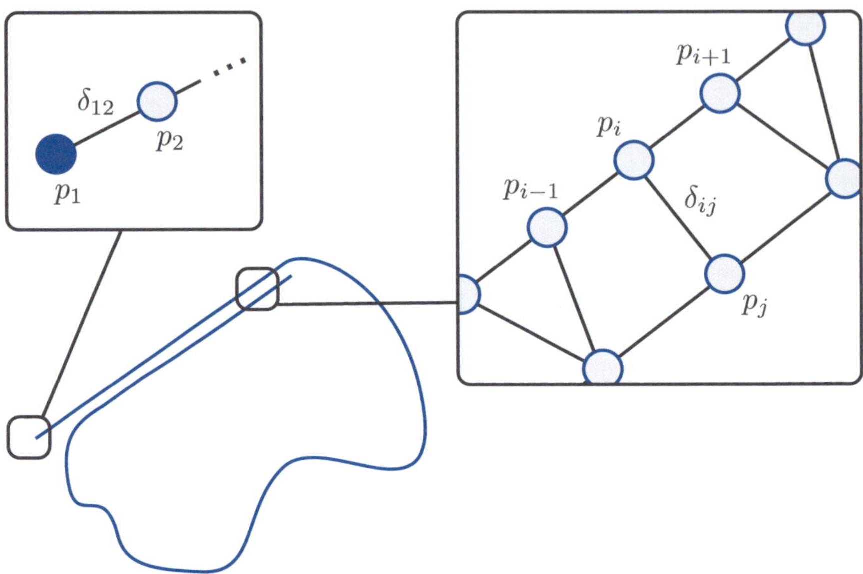

Figure 4.1: The factor graph of the pose optimization is shown. All poses except the first are optimized such that their pairwise relative distances correspond to the visually estimated motion δ_{ij} as good as possible.

4.2   Loop Closure Detection

A crucial ingredient of the pose graph optimization above is the detection of pairs of stereo images that show the same place. It is important to enforce map consistency in areas of self overlap which cannot be achieved without such detection. The goal is to find pairs of stereo images s_i and s_j that show the same location even though the images are captured on different times, possibly days. We present our approach for place recognition in this section.

First one single feature vector is computed for every image of $\mathcal{S}$ and pairwise similarities are computed from them. This matrix of pairwise similarities is thereafter refined by a post processing step to achieve robustness to visual aliasing and ambiguities.

In the sequel we refer to the feature vector that describes an image as *holistic feature vector* [66]. The left image of a stereo frame is sampled down and partitioned into 4×4 equally sized tiles each of which is 48×48 pixels in size. Then, one DIRD descriptor is computed for the center part of each tile. All sixteen DIRD features of one image are concatenated to form the holistic feature vector. The

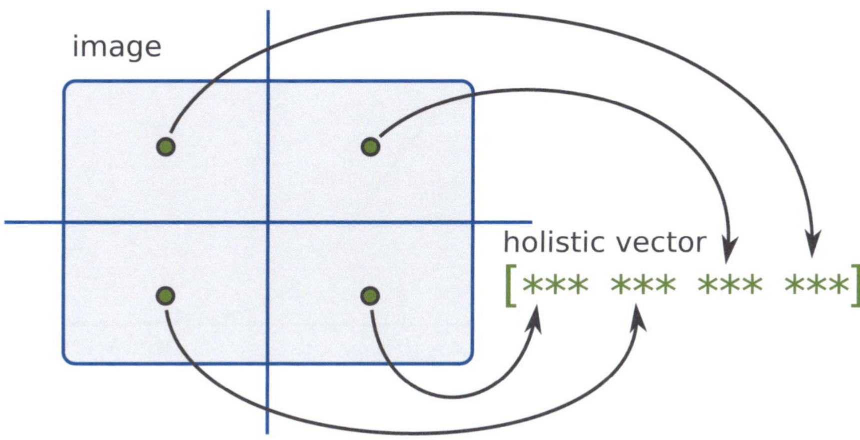

Figure 4.2: A holistic feature is computed for each image. To this end the image is sampled down and tiled into $n \times n$ equally sized tiles. Each tile is described by one DIRD [36] vector and these vectors are finally concatenated. The example shows a 2×2 tiling whereas a 4×4 tiling is used in the experiments.

final holistic vectors are of dimension 3456 where each element of the vector is a single byte. The process is schematically depicted in Figure 4.2. Note that these holistic features are also stored as part of the final map (see Section 4.4).

Now, let $f_1, \ldots, f_N$ be all holistic features of the map with f_i extracted from the left image of s_i. A column vector D_i of L1 distances is then computed by

$$D_i = (\|f_1 - f_i\|_1, \ldots, \|f_N - f_i\|_1)^T .$$

(4.11)

for every index $i \in \{1, \ldots, N\}$. Note that this operation can be performed quite efficiently on modern CPUs using SIMD (single instruction, multiple data) instructions since DIRD features are byte vectors.

Simply taking the minimizing argument of (4.11) as the result of the place recognizer is error prone and susceptible to visual aliasing and ambiguities. Hence, we introduce a post processing step to refine the search next. The idea is to expect that for a correct match of images s_i and s_j their holistic features f_i and f_j are similar (in a vector distance sense). Moreover we also expect neighboring poses of s_i and s_j to match as well. For this correct match to be accepted we also require $f_{i'}$ to match $f_{j'}$ if pose p_i is close to $p_{i'}$ and p_j is close to $p_{j'}$. Hence, we search for subsequences of features that match.

We formalize this requirement by first defining the similarity column vector

$$S_i = (\text{logit}(\|f_1 - f_i\|_1), \ldots, \text{logit}(\|f_N - f_i\|_1))^T$$

(4.12)

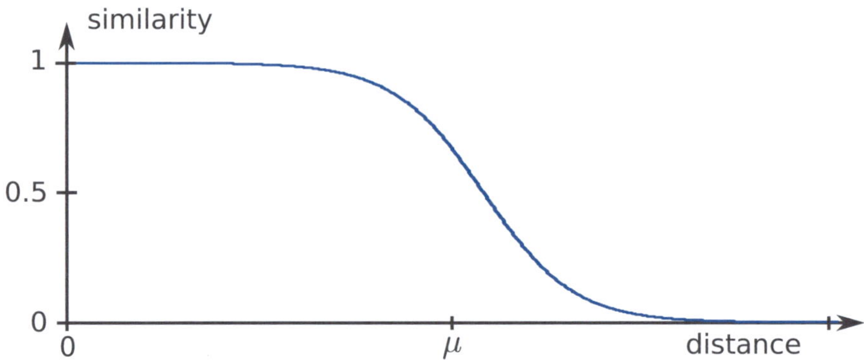

Figure 4.3: The logistic function $\mathrm{logit}(\cdot)$ that translates all distance values into similarity scores between zero and one is depicted.

with $\mathrm{logit}(\cdot)$ being a logistic function which translates all distances of (4.11) into similarity scores in the range $(0,1)$. The logistic function is parameterized such that it translates large distance values to a similarity score of (near) zero and small distances to one. Values in the range of μ which is half of the expected maximum value result in similarity scores near 0.5. Hence, the logistic function is

$$\mathrm{logit}(d) = \frac{1}{1 + \exp\left(\frac{-(d-\mu)}{\sigma}\right)} \tag{4.13}$$

with d being the distance value and σ an appropriate scaling value which determines the slope of the transition range near μ. The $\mathrm{logit}(\cdot)$ function is plotted in Figure 4.3. Next the similarity matrix

$$S = (S_1, \ldots, S_i) \tag{4.14}$$

with column vectors of (4.12) is defined. Thus, $S_{j,i} \in S$ denotes great similarity of the image s_i to the image s_j if its value approaches one. A such defined similarity matrix is best shown visually and an example is depicted in Figure 4.4 (left). An off-diagonal streak of high similarities can clearly be seen. We aim at finding such streaks which hint at well matched subsequences.

For any image s_j that we consider to show the same place as image s_i we search for a streak as in Figure 4.4 that ends at row j in column i. Formally let

$$T_{j,i} = \max_{\substack{j_1,\ldots,j_L \text{ s.t.} \\ (j_{k-1}-j_k)\in\{0,1,2,3\} \\ \text{and } j_L=j}} \sum_{k=1}^{L} S_{j_k,i-k+1} \tag{4.15}$$

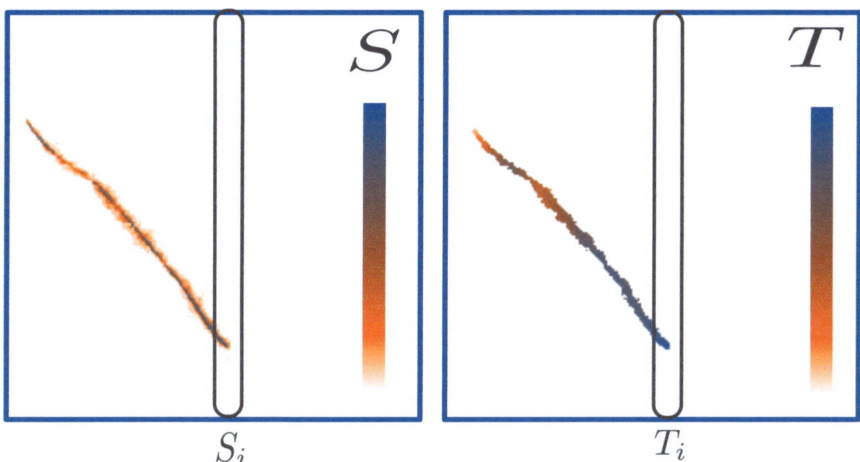

Figure 4.4: Sample matrices S and T are shown. S shows pairwise similarities between poses of the survey trajectory. The color scale is shown from orange (low) to blue (high). The streak indicates that two subsequences overlap and show the same area. The search for such streaks is performed by dynamic programming and the result is shown on the right. The right matrix T shows an increase in certainty into the loop closure (ranging from orange (low) to blue (high)). The main diagonal is not shown.

be the maximum sum of one such streak of length L. The matrix T with elements $T_{j,i}$ can be computed efficiently from S by dynamic programming and one example is shown in Figure 4.4. We compute $T_{j,i}$ only for those j, i which seem promising which means that $S_{j,i}$ exceeds a conservative threshold. Note that S and T are both extremely sparse and memory efficient. The streak length in our experiments is $L = 30$.

The last step of place recognition is to suppress locally non-maximal place matches. The right of Figure 4.4 shows that a few places are matched within a small neighborhood. For pose graph optimization and all subsequent steps a slightly sparser graph is beneficial. Hence, we keep only the maximum match in each column. The non-maxima suppressed matrix is denoted by U. It contains the values $U_{j,i} = T_{j,i}$ if j, i is locally maximal. Finally, we accept all image pairs j, i as place matches if the value $U_{j,i}$ exceeds the threshold $\tau = \frac{1}{3}L$.

We note that the method presented above is unable to detect a previously visited place when observed from an opposite driving direction. An algorithmically trivial solution to largely mitigate this effect is to record imagery for a forward and backward facing camera simultaneously. The distance computation in (4.11) and

(4.12) can be replaced by a simple surrogate

$$\min \left\{ ||f_i^f - f_j^f||_1 + ||f_i^b - f_j^b||_1, ||f_i^f - f_j^b||_1 + ||f_i^b - f_j^f||_1 \right\} \qquad (4.16)$$

with f^f and f^b being the forward and backward feature vectors.

4.3 Landmark Estimation

Next, we elaborate the computation of the set of landmarks which are used during localization. We associate salient image points across all images of the sequence $\mathcal{S}$. We refer to a set of pixel positions belonging to a single point in 3D as a tracklet. Every landmark that is finally stored in the map is computed from exactly one tracklet. It remains to show how to compute the 3D position of one landmark $l_j \in \mathbb{R}^3$ from its tracklet. Recall that a stereo setup is used for mapping. Thus, the pixel position and disparity of l_j is available when observed from a set of poses $p_k, \ldots, p_{k+K}$. We summarize pixel positions and disparities in the measurement vectors $z_k = (u_k, v_k, d_k)^T, \ldots, z_{k+K} = (u_{k+K}, v_{k+K}, d_{k+K})^T$ for the landmark of interest. Finally, the error function

$$E_{\mathrm{lm}}(l_j) = \sum_{\kappa=k}^{k+K} ||\pi(l_j, p_\kappa) - z_\kappa||^2 \qquad (4.17)$$

provides a goodness of fit of l_j with respect to the measured pixel positions and the poses $p_k, \ldots, p_{k+K}$ that are fixed (see Section 4.3). The function $\pi(l, p)$ computes pixel position and disparity for a landmark l observed from pose p [29]. The 3D landmark position can be estimated by

$$\hat{l}_j = \arg\min_{l_j} \left\{ E_{\mathrm{lm}}(l_j) \right\} \qquad (4.18)$$

and is found by NLS estimation which is repeated for every tracklet.

The factor graph of (4.17) is shown in Figure 4.5. It shows the poses $p_k, \ldots, p_{k+K}$ and the landmark l_j. Poses are solid which denote a fixed variables (ones that are not optimized). The landmark is shown by a hollow stars indicating a variable which is optimized and is exactly the argument of (4.17). The summation of (4.17) extends over all edges of the graph each of which corresponds to a pixel position and disparity. Hence, we estimate the landmark positions for fixed poses. This allows to compute each landmark independently of all other ones which can easily be done in parallel. Moreover, the computation time scales only

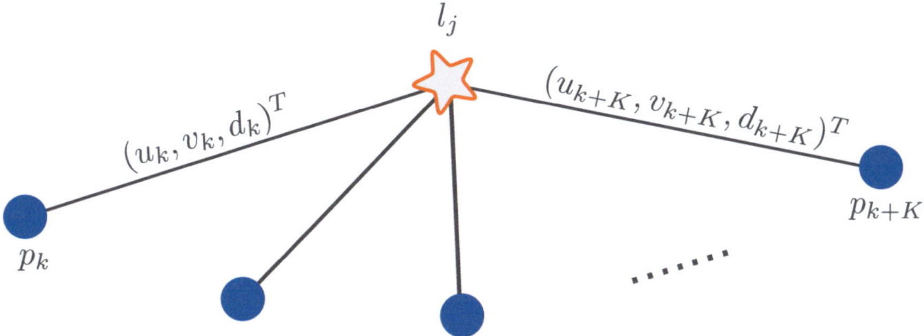

Figure 4.5: The factor graph of the landmark estimation is shown. The landmark l_j is observed from multiple pose $p_k, \ldots, p_{k+K}$. Poses are kept fixed during landmark estimation and not subjected to optimization.

linearly with the map size as opposed to squared in the case of joint estimation of poses and landmarks.

Using (4.18) for every tracklet/landmark yields the 3D position of every landmark. Finally, we prune some landmarks from the map that seem inappropriate for localization. Back projection errors and lengths of the tracklets are heuristically thresholded for this purpose. For robust and reliable landmark association during online localization, landmark feature vectors are computed. In our case we use our illumination robust yet efficient novel DIRD [36] descriptor. For the example landmark l_j (see Figure 4.5) one descriptor is computed for every pose $p_k, \ldots, p_{k+K}$ it is observed from and stored within the map data structure (see Section 4.4).

4.4 Map Data Structure

The final visual map consists of all poses $p_1, \ldots, p_N$ their holistic features $f_1, \ldots, f_N$, all landmarks $l_1, \ldots, l_M$ and their visual description. For fast online retrieval we store all landmarks visible from a given pose p_i and their feature vectors extracted from that particular image i together in one file. Hence, every landmark is represented by a multitude of landmark feature vectors; one for each image the landmark was observed from. This wasteful appearing over parameterization, however, contributes much to a reliable association during online operation. It frees us from any struggle related to scale and/or rotation invariance. We simply match landmark features of the nearest pose. Moreover, the search for potentially

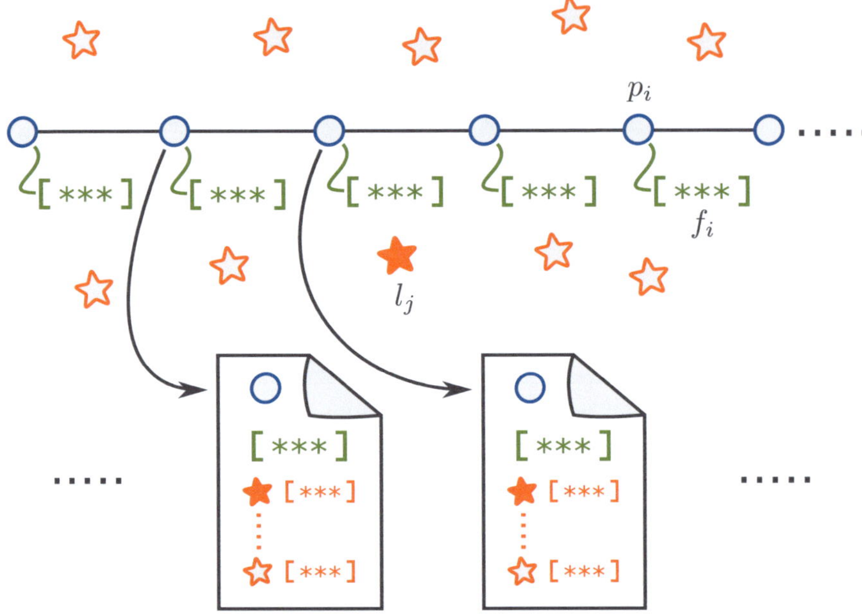

Figure 4.6: The map data structure is shown. For every pose of the map, one single file is created. The file contains the estimated pose, the holistic feature describing the pose appearance and all landmarks visible from this pose. A sample landmark is shown solidly. It is stored in both files with the visual descriptor computed independently for each pose.

matchable landmarks is easy. Only landmarks that are stored for the currently nearest mapping pose are used. The map data structure is depicted in Figure 4.6.

Some cases require a global position estimate during localization. An example is a self driving vehicle which shall follow a certain trajectory. This trajectory may have been recorded by a high precision GNSS receiver or may have been automatically created from digital road maps. Then, this trajectory can only be followed by the vehicle if the localization estimate is expressed in the same coordinate system (e.g. latitude, longitude, altitude). Hence, we propose to slightly extend the above algorithm to obtain a global reference. This step requires a GNSS receiver during mapping and is optional and only mentioned for completeness.

Suppose there exists a global position g_i for every stereo frame s_i of $\mathcal{S}$. Then we simply set $p_i = g_i$ when storing the poses of the map. Note, however, that the global position is only used when saving the map to disk. In particular a visual odometry induced pose estimate is used for the creation of the landmark positions.

Landmark positions are stored relative to the poses they are observed from and their estimate it thus not effected by this pose update. Thereby the pose estimates can be altered (e.g. set to a global position) after the map computation is already completed. Moreover any slight map/trajectory updates to enforce consistency with other map data is easily achieved and the entire map does not need to be completely recomputed.

Finally, we note that using a global pose estimate for the estimation of landmark positions is infeasible. GPS position estimates lack the required precision too often. The problem becomes especially pronounced when sudden GPS jumps occur. This happens frequently in practise when e.g. leaving street canyons or the like. In the areas of GPS jumps no landmark positions can be estimated that explain the observed pixel positions well enough and all landmarks are classified as outliers. Hence, these areas would become essentially landmark free which is a truly undesirable property.

4.5   Experiments

We present the result of a large scale mapping experiment next. Several sequences of stereo images are recorded and no GPS is used anywhere. The focus of the first set of experiments lies on the proper recognition of correct place matches. Thereafter, we present the results of the landmark estimation process.

First, we test the place recognizer of Section 4.2. We recorded a sequence of 14500 stereo images. For the following place recognition experiments we have recorded imagery for a forward and backward facing stereo setup to detect places from different driving directions. Both stereo setups are only used for place recognition and all subsequent steps rely on only one of the setups.

Figure 4.7 shows the similarity matrix S and its post processed version U. The size of the matrix is quite large and some example areas are shown with more appropriate zoom. One loop closure (first zoom picture) occurs in an area of visual ambiguity. From the simple vector representation of an image no reliable place recognition can be achieved. The post processed (and non-maxima suppressed) similarity matrix U however shows good noise removal properties.

The result of pose graph optimization is shown in Figure 4.8. The left part shows the result of graph optimization where the place recognition step has been skipped. The graph contains no loop closures and the resulting estimate is a mere motion integration. The inevitable drift that occurs in this situation is clearly visible. The graph shows great inconsistencies.

The right part of Figure 4.8 shows the result of pose graph optimization once the result of place recognition is incorporated into the mapping process. Areas of de-

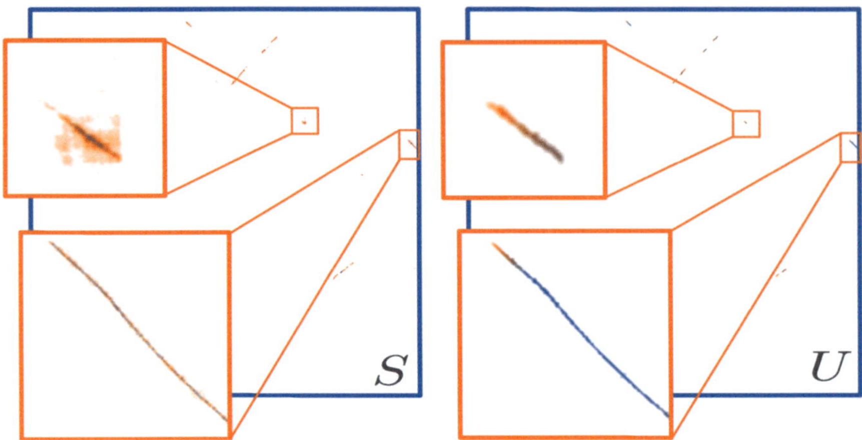

Figure 4.7: The similarity matrix for the first test set is shown. The dynamic programming procedure and non-maxima suppression produce a noise removed version depicted on the right side.

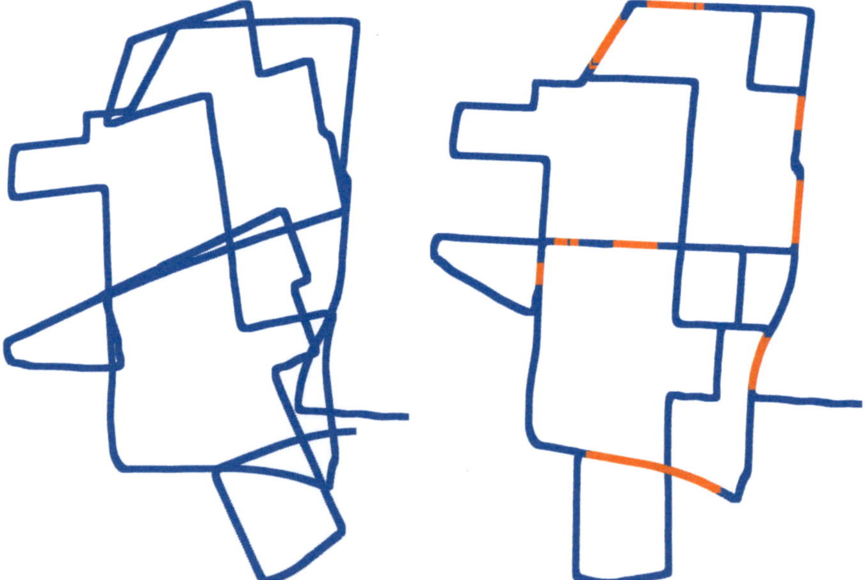

Figure 4.8: The estimated pose graphs for test set one are shown. The left part shows the loop closure free version whereas the right shows the final result after integrating the detected loops which are marked orange.

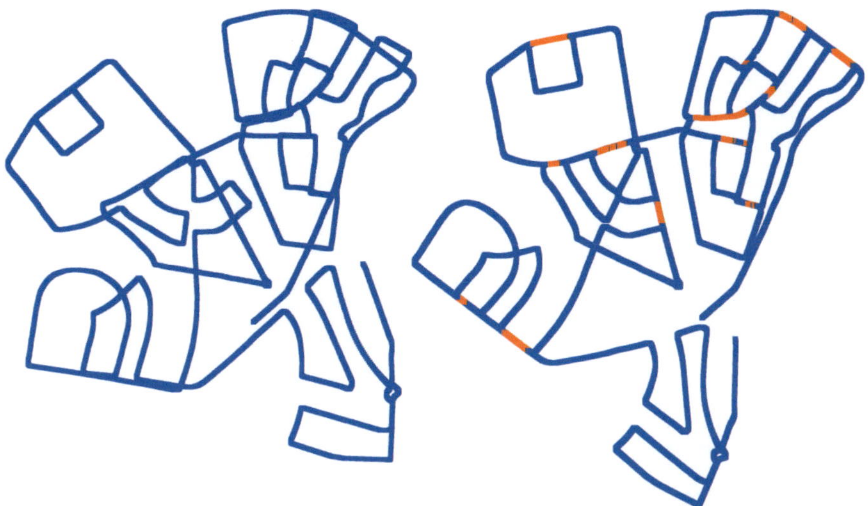

Figure 4.9: The estimated pose graphs for test set two are shown. The left part shows the loop closure free version whereas the right shows the final result after integrating the detected loops. The graph contains 25000 poses. Orange indicates overlap again.

tected self overlap are highlighted in orange. Each area of self overlap corresponds to exactly one streak of the matrix U in Figure 4.7. All inconsistencies are clearly resolved. We note that the pose graph optimizer is a non-robust version since the place recognizer has not introduced a single false positive link into the pose graph. Hence we refrained from integrating any outlier detection schemes as e.g. in [66]. The same holds for all following experiments. Any false loop closure detection would result in a severely and clearly visibly distorted pose graph estimate.

To assess the robustness of the presented method we have repeated this experiment for a longer image sequence of 25000 images. It was recorded in a small suburban area and covers almost all streets of that suburb. Trying to record imagery for all streets naturally requires loop rich trajectories. Some streets have been traveled several times during this survey. The resulting graph contains various loops of different sizes.

The result of place recognition is shown in Figure 4.9. First the pose graph without any place recognition is shown on the left. Heavy inconsistency and drift can be seen. The right part shows the pose graph after place recognition. Areas of self overlap are again denoted by orange.

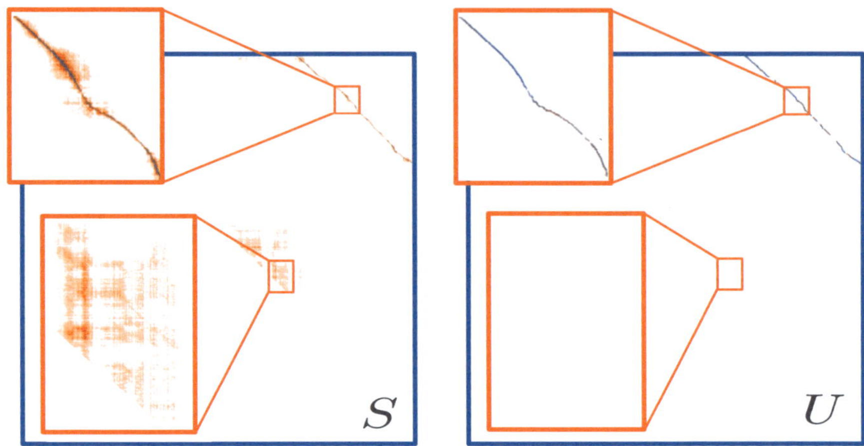

Figure 4.10: The similarity matrix for the large loop test set is shown. Noise due to visual aliasing is reliably removed.

The two aforementioned experiments demonstrate that places are reliable detected in suburban areas. To evaluate the place recognizer once faced with large loops we have recorded an image sequence of a single large loop on a two lane major street. The similarity matrix S and its post processed version U are depicted in Figure 4.10. The similarity matrix shows areas of high self similarity. This is caused by a large alley like appearance which only slowly changes. The dynamic post processing, however, is able to reliably detect it and remove it from the final place recognition matrix U shown on the right of Figure 4.10. The area of overlap manifests as a continuous, off-diagonal, long streak and can clearly be seen in both S and U.

The resulting pose graph estimate is shown in Figure 4.11. The left part shows the loop closure agnostic version whereas the right shows the final pose graph estimate after the integration of the place recognition. The orange (detected self overlap) is not continuous. In this experiment we have found that lane changes between traversals cannot be reliably handled by our method and we attribute the discontinuous detection to that. The holistic appearance of a place changes substantially once observed from a lateral offset of several meters. We believe that this drawback is inherent to the approach and can only be overcome by more sophisticated methods. We propose an extension in the last chapter 7 and simply accept it for now. Nevertheless, the result is satisfactory. Missing some spots of self overlap

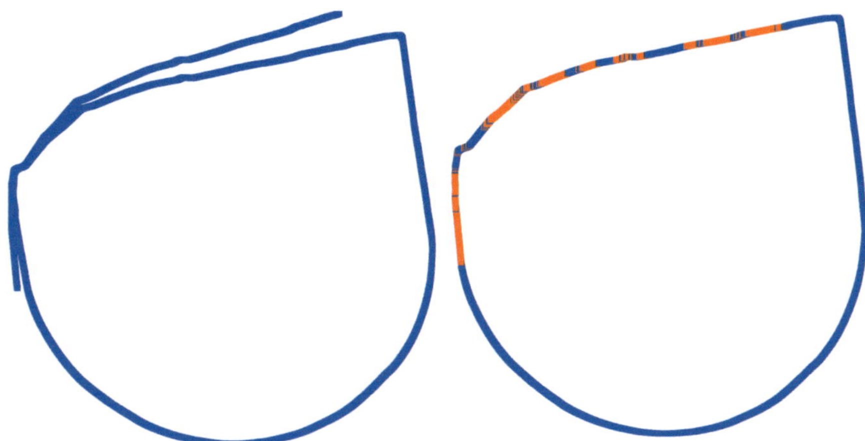

Figure 4.11: The estimated pose graphs (with and without loop closure) for the large loop data set are shown.

is far more favorable than introducing false loop closures. This, however, did not happen in any of the experiments.

Finally, we stress test the place recognizer by a large scale experiment. We recorded a fourth stereo sequence whose trajectory overlaps with each of the three preceding ones. We then ran the place recognizer and pose graph optimizer on the joint set of all recordings. The result is shown in Figure 4.12. The fourth recording (glue trajectory) is depicted in orange whereas the first three are shown in blue. All areas of cross and self overlap are detected correctly. The final pose graph optimizer does not handle outliers and assumes all links to be correct. However, this is essentially superfluous since no false detections are output by the place recognizer. Next, we show some examples of the landmark estimation process for a backward facing stereo camera setup. Figure 4.13 shows a typical image of one of the sequences. The orange dots and circles of the image indicate the pixel positions for which a landmark was computed. In fact, each dot is a small circle and the circle radius denotes the back projection error which we seek to minimize during mapping. A larger circle therefore indicates a worse landmark position estimate. All circles shown in the figure are stored in the map. In particular those landmarks that are pruned after estimation due to short tracklet lengths or bad back projection errors are not shown. The following car for instance does not contain any landmarks even though it lends itself nicely to feature extraction. Moving objects are easily handled by our pruning heuristic.

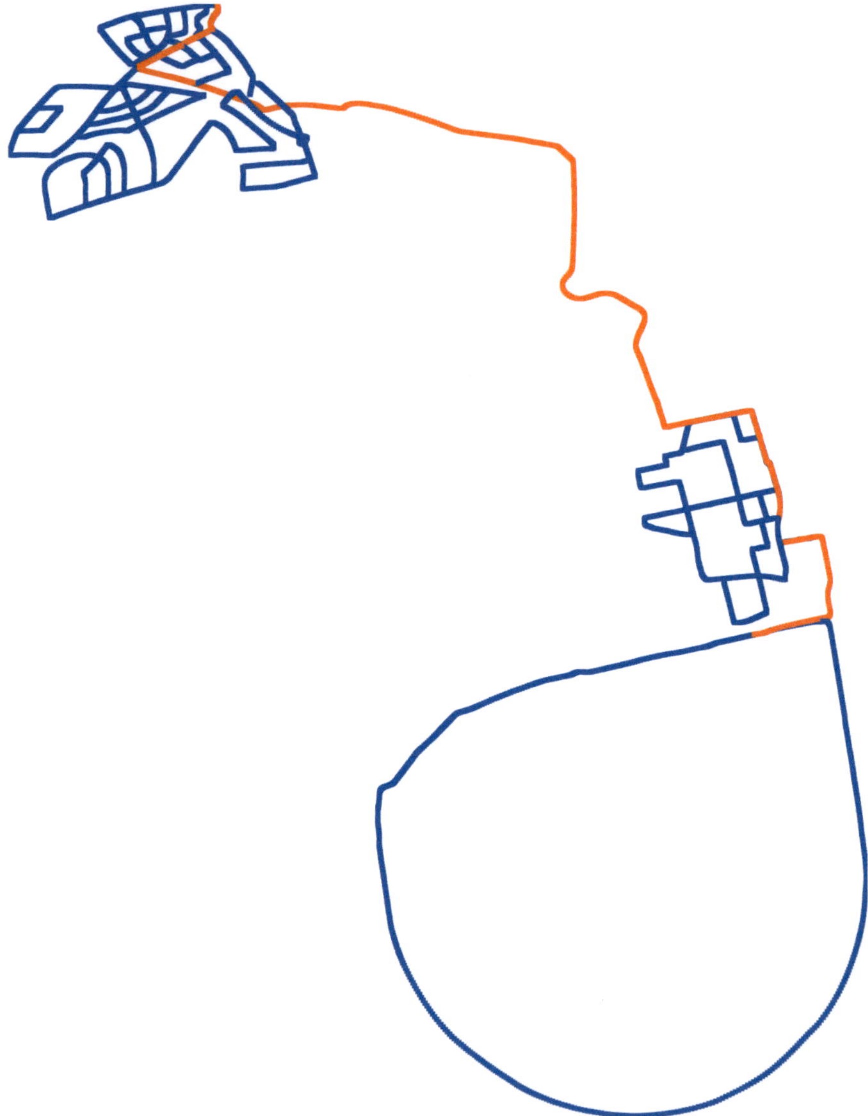

Figure 4.12: The place recognizer is stress tested by merging all aforementioned trajectories by one "glue" trajectory shown in orange. All areas of self and cross overlap are correctly detected.

Figure 4.13: The pixel positions for which a landmark was estimated are shown for a backward facing camera configuration. All features on the car are correctly identified as non-stationary and are therefore not stored in the map.

Figure 4.14 shows another image with detected landmarks. This image was captured in a rural area. One can observe that much fewer landmarks are extracted in that case. Roughly half of those landmarks are located on vegetation which is not persistent and may vary over larger time spans. Another significant portion is located far behind the vehicle close to the horizon which cannot contribute much to a localization estimate. We identified this drawback of "feature poverty" in rural areas as the greatest weakness of the entire mapping/localization approach. Localization robustness and accuracy is higher in feature dense urban areas. Nevertheless, we note that localization is accurately possible in most rural cases if there exists at least a modicum of structure. Moreover, we see much room for improvement and propose an extension for landmark extraction in the last chapter 7 which is specifically tailored to road surface features which also exist in rural areas.

Figure 4.15 shows several examples of small map chunks viewed from the birds eye perspective. Following the color scheme, the poses of the mapping trajectory are shown in blue whereas the landmark positions are shown in orange. For better visibility only those landmarks which are close to the mapping trajectory are shown. Landmarks much further away are neglected in that figure even though they are stored in the map data structure. Parked cars, house facades and crossing roads can be found in this view.

Figure 4.14: Another image where landmark pixel positions are highlighted. Rural areas contain much fewer landmarks than urban areas. Many features are far away from the ego vehicle and do not contribute much to a ego pose estimate during localization.

Finally we state some statistics related to mapping. The number of landmarks that are detected in one image of the sequence ranges from one hundred to three thousand depending on both the area (rural, urban) and the traveling speed. The considerably high average landmark density coupled with the approach of storing landmarks and their visual features redundantly leads to an average map storage size of almost one gigabyte per kilometer.

The visual odometry which is used to create the pose graph runs with approximately 10Hz on modest computing hardware. The time for holistic feature extraction is negligible (approx. 5ms per image). The place recognizer scales quadratically with the number of poses and takes about thirty minutes for a pose graph of 50000 poses where almost all time is spend on the similarity matrix computation S. The dynamic programming refinement takes a few minutes. The time for pose graph optimization varies considerably and depends on the number of poses and even more so on the topology of the graph. By far the most time is spend on landmark feature extraction. Their 3D estimation is very efficient and is negligible compared to the feature association step. For feature extraction, association and position estimation we achieve a speed of approximately three kilometers in one hour. However, the entire mapping tool chain runs fully automatically and requires no manual intervention. Moreover, it is embarrassingly parallel and lends itself to multi core/computer architectures which should allow massive speed ups.

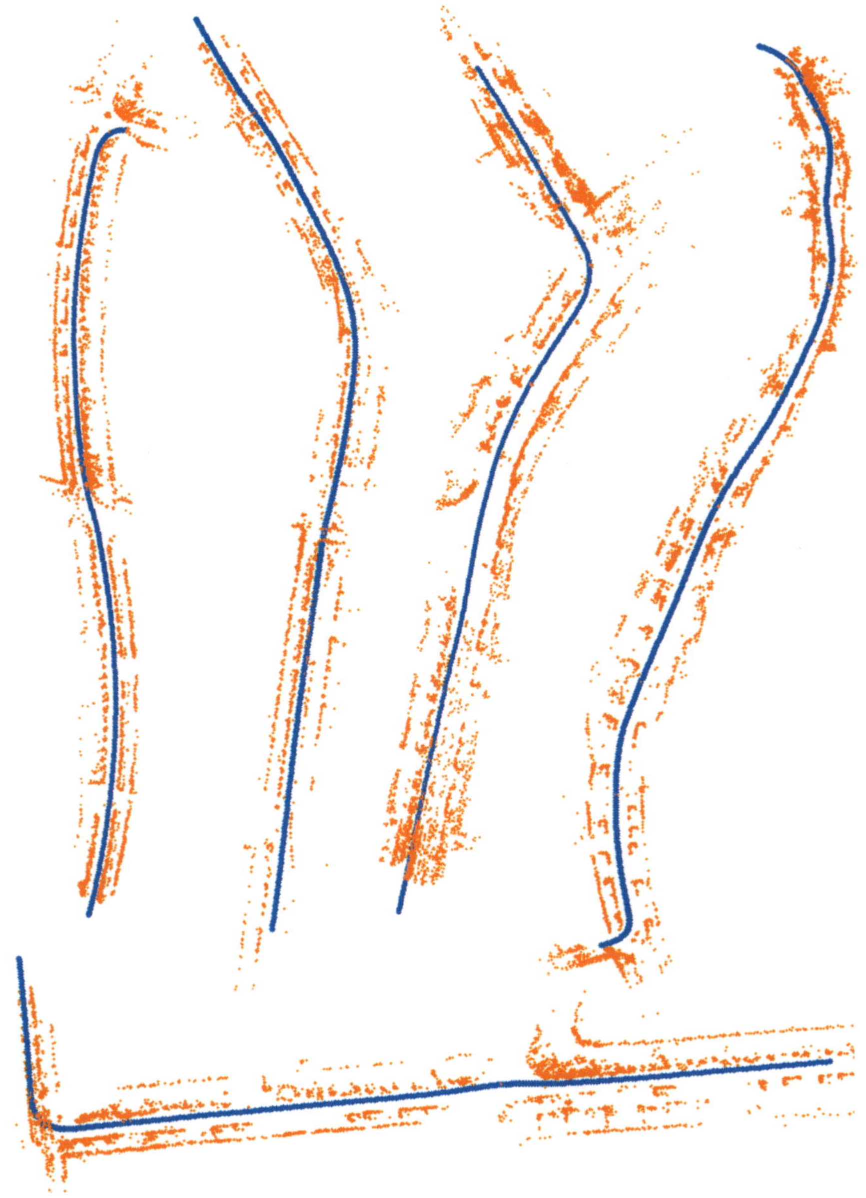

Figure 4.15: The top view of some areas of the map are shown. Blue denotes the estimated trajectory and landmarks are depicted in orange.

5 Localization

Next, we present the localization algorithm. A single monocular camera is used and we show how to localize that camera relative to the visual map as described in Chapter 4. First, a rough overview is presented. At that point we spare the details before elaborating the technicalities in Sections 5.1 and 5.2 respectively. Earlier versions of this approach have appeared in [42, 41]. Figure 5.1 shows an overview of the algorithm. We distinguish three states the localizer may be in.

On start up nothing is known about the camera position. In fact it is not even known whether the camera is anywhere inside the area for which a map is available. Hence, the method is in state unknown. In that state the goal is to find the pose of the map that is nearest to the current ego position. This is achieved by extracting one holistic feature for the current camera image and match that against

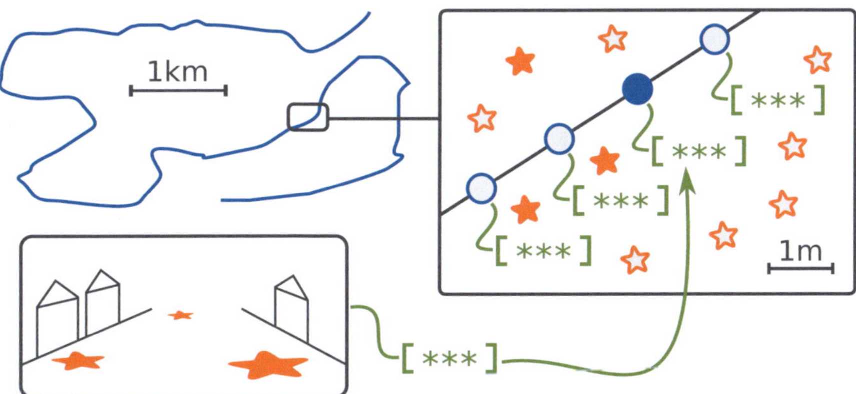

Figure 5.1: An overview of the localization algorithm is shown. The map consists of poses (blue circle), associated holistic features (green brackets with asterisk) and landmarks (orange stars). The current camera image is depicted on the bottom. A holistic feature is extracted from it and the most similar vector of the map is found (arrow). All landmarks of the vicinity are then associated with salient points of the image. The three solid landmarks (stars) are associated successfully in the example and shown in the camera image. The landmark associations are translated into a high precision metric ego pose estimate.

the entire set of holistic features of the map. Figure 5.1 exemplary shows the current camera image on the bottom. Holistic features are denoted by green brackets with asterisks. The nearest pose is highlighted by a solid circle in that example. We refer to this type of localization as topological localization and it is closely related to the place recognizer of Chapter 4. Details of how to develop the place recognizer into a light weight topological localizer is elucidated detailedly in Section 5.1. Once a nearest pose is found the localizer changes state from `unknown` to `topological`.

A topological localization yields the nearest pose of the map which forms the starting point for metric localization. Thus, the aim is to change state from `topological` to `metric`. To this end, landmarks of the vicinity of the nearest pose are loaded from disk and associated with salient points of the current camera image. The current camera image at the bottom of Figure 5.1 shows three example landmarks highlighted by solid orange stars which can also be found in the top view of the map directly above. Pixel positions of the landmarks whose 3D world coordinates are known can finally be used to estimate a metric ego pose. Details of this step are spelled out in Section 5.2. If this metric localization has succeeded for a few time steps the localizer changes state from `topological` to `metric`. Note that the search for the nearest mapping pose is straight forward when in state `metric` and obviates the need for global searching. When leaving the map the internal state changes back to `unknown`.

5.1   Topological Localization

The goal of topological localization is to find the pose of the map that is closest to the current ego pose. If this pose is known, all relevant landmark positions and their visual descriptors can be loaded from disk and used to compute a metric ego pose estimate (see Section 5.2). Since the map size can become considerably large we herein present a method for efficiently finding the nearest neighbor pose. To this end, we slightly modify the place recognizer of Section 4.2 and derive a light weight topological localizer. The current camera image is used to extract one holistic feature vector from it. That feature vector is thereafter compared to each of the holistic feature vectors of the visual map. A dynamic programming procedure is used to achieve robustness in areas of visual ambiguity and aliasing.

The current camera image is down sampled and partitioned into 4×4 equally sized square tiles each of which is 48×48 pixel in size. The center pixel of each tile is described by one DIRD vector. These sixteen feature vectors are concatenated to form one holistic vector and is denoted g_i for the current camera image with i being the current time index. Recall that the holistic feature vectors of the map are stored in the map data structure. Let these map feature vectors be $f_1, \ldots, f_N$ for the N poses of the mapping trajectory. The search for the spatially nearest pose now corresponds to the search of the pose of the map that is closest in appearance hence in the space of holistic feature vectors. Since a simple nearest neighbor search is susceptible to visual aliasing we apply the dynamic programming refinement of the place recognition of Section 4.2.

Let $S \in \mathbb{R}^{N \times i}$ be a similarity matrix with columns

$$S_i = (\mathrm{logit}(\|f_1 - g_i\|_1), \ldots, \mathrm{logit}(\|f_N - g_i\|_1))^T \in \mathbb{R}^N \tag{5.1}$$

with $\mathrm{logit}(\cdot)$ being a logistic function which translates all L1 distances into a similarity scores in the range $(0,1)$. The $\mathrm{logit}(\cdot)$ function is defined by

$$\mathrm{logit}(d) = \frac{1}{1 + \exp\left(\frac{-(d-\mu)}{\sigma}\right)} \tag{5.2}$$

with distance value d and shift parameter μ and scaling factor σ (cf. (4.13) and Figure 4.3). Hence the similarity matrix is defined as

$$S = (S_1, \ldots, S_i) \tag{5.3}$$

with column vectors of above. If an element $S_{j,i}$ of the matrix is approaching one then it denotes a high visual similarity of the current pose i to the map pose j. However, to accept j as the nearest pose we also expect the immediately preceding online feature g_{i-1} to show high similarity to f_{j-1} of the map. Since traveling velocities between mapping and localization may be different we alternatively allow that f_{j-3} matches g_{i-1} for example. In fact, we search for a short subsequence of L features $g_{j_1}, \ldots, g_{j_L}$ of the map that matches the past L online features $f_{i-L}, \ldots, f_i$ while constraining $(j_{k-1} - j_k) \in \{0,1,2,3\}$. For a potential nearest pose j we define

$$T_{j,i} = \max_{\substack{j_1, \ldots, j_L \text{ s.t.} \\ (j_{k-1} - j_k) \in \{0,1,2,3\} \\ \text{and } j_L = j}} \sum_{k=1}^{L} S_{j_k, i-k+1} \tag{5.4}$$

to be the maximum match of a subsequence of length L. For efficiency reasons we compute the column vector T_i for the current time index only for those j that seem promising enough which is $S_{j,i} > \tau_S$. Moreover, we allow to compute at most M values $T_{j,i}$ at each time step. We sort all $S_{j,i} > \tau_S$ and choose the M greatest ones. Finally, a nearest pose j is accepted if $T_{j,i} > \tau_T$. In all experiments we have set the length of the subsequence $L = 30$, the number of candidates $M = 10$ and the threshold $\tau_T = \frac{1}{3}L$.

Computing the row of similarities S_i takes time linear in the number of mapping poses whereas the refinement of computing T_i is constant. Nevertheless, the similarity computation can be implemented very efficiently on modern CPUs using SIMD instructions since DIRD features are byte vectors.

Note that computing visual similarity for topological localization as presented above is concise and globally optimal. We see some advantages over filter based alternatives like particle or histogram filters. Unsolved problems like the correct number of particles or issues related to particle depletion do not arise. Moreover, the topological localization is very efficient and takes only a very few milliseconds on a single core for moderate map size (few ten thousand features). The linear complexity of the similarity computation S_i will eventually cause a loss of real time capability for very large map sizes. A reduction of the dimensionality of the holistic features seems an obvious step to mitigate (but not avoid) this issue. More thoughts on the problem and potential solutions are discussed in chapter 7. For all map sizes of the experiments we have not experienced any run time problems. Topological localization is very efficient and takes up only a small fraction of the total computation time.

5.2 Metric Localization

Our metric localization method follows a two step approach. First an initial (metric) estimate of the camera pose is computed from landmark associations. Henceforth, we will refer to this as a one-shot estimate. Since these one-shot estimates can have varying accuracy and may even fail in some very unfavorable situations a sliding window history of one-shot estimates is stored. During the second step, these past one-shot estimates are re-optimized jointly. To this end, we fit a motion model to the sliding window of past one-shot estimates. Both steps are elaborated in greater depth next.

During metric localization it is assumed that the pose of the map that is closest to the current camera position is already known. Either it is easily inferred from the immediately preceding time step or the topological localization provides a hint. This knowledge allows to load the associated map pose file which contains all nearby landmark positions and their visual descriptors from disk (see also Figure 4.6). Next, these landmarks are associated with pixel positions of the current camera image. Salient points are extracted, described by DIRD and matched with those of the map. The search space within the image plane can be restricted quite heavily since a good ego pose estimate is known from the previous time step already.

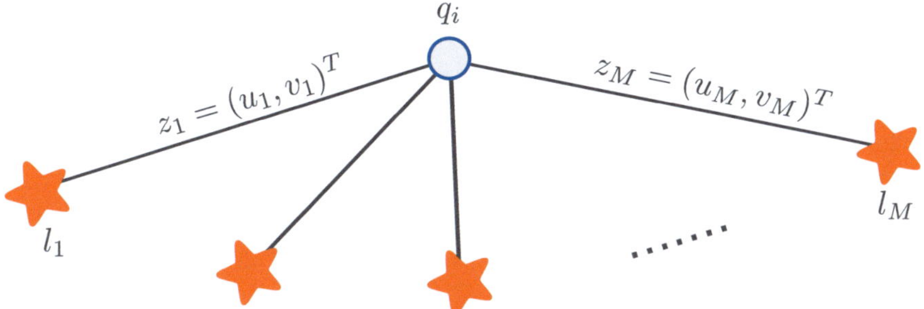

Figure 5.2: The factor graph of the NLS problem of (5.5) is shown. The current pose (blue circle) is optimized for fixed landmark positions (solid orange stars) such that it matches the measured pixel positions z_m best.

Let the set of landmarks successfully matched be $l_1, \ldots, l_M$ and their associated pixel positions be $z_1 = (u_1, v_1)^T, \ldots, z_M = (u_M, v_M)^T$. The current ego pose is denoted by $q_i \in SE(3)$.

The 3D landmark positions are kept fixed and a one-shot estimate $\bar{q}_i$ is found by seeking the minimizing argument of

$$E_{\text{one}}(q_i) = \sum_{m=1}^{M} ||\pi(l_m, q_i) - z_m||^2 \tag{5.5}$$

where $\pi(l, q)$ computes the pixel position of the 3D point l projected into the camera at pose q [29]. This one-shot estimate is found by NLS estimates and denoted by

$$\bar{q}_i = \arg\min_{q_i} \{E_{\text{one}}(q_i)\}. \tag{5.6}$$

The factor graph associated with the NLS problem (5.5) is visualized in Figure 5.2. Landmarks are denoted by stars and since they are kept fixed (are not optimized for) in (5.5) are depicted with solid colors. The pose q_i is denoted by a hollow circle and is the argument of (5.5). The summation of (5.5) extends over the edges of the graph which are labeled with the measured pixel positions z_m.

Since (5.5) is a quadratic error function it is naturally very susceptible to any outliers. Outliers can arise from miss-associations which cannot be fully avoided in practice despite a carefully designed feature matcher. Moreover, any incorrectly estimated landmark can cause such outliers as well. Undetected outliers can cause catastrophic divergences of the pose estimator. Therefore, we wrap the estimate of

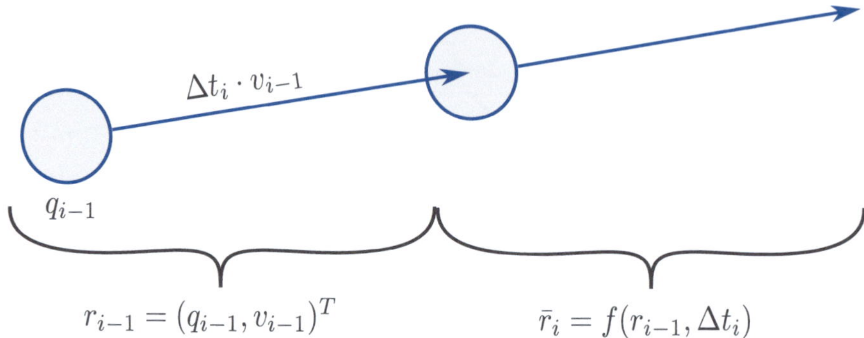

Figure 5.3: The prediction of a pose from a previous *velocity augmented pose* is shown. The pose contains a velocity vector $v = (v_x, v_y, v_z)^T$ which can be used to extrapolate with the known time lag Δt.

(5.6) in a random sampling consensus [24] (RANSAC). We randomly draw minimum sets of three landmarks, estimate $\bar{q}_i$ and evaluate all landmark pixel positions for support of the current hypothesis. After one hundred such iterations the largest inlier set is optimized jointly in a NLS sense to yield the final one-shot estimate $\bar{q}_i$. Measurement covariance matrices have been neglected so far in favor of better readability. The norm of (5.5) is in fact a squared Mahalanobis norm which considers measurement uncertainty. Next, we introduce the pose adjustment step of the localization algorithm which jointly re-optimizes a set of past one-shot estimates. This step, however, requires a certainty measure of the estimate $\bar{q}_i$ and we denote its covariance matrix by Σ_i. We find a judicious choice of Σ_i by checking the number of inlier landmark associations (according to RANSAC). Uncertainty is increased for fewer inlier landmark matches and vice versa.

A sliding window history of past one-shot estimates $\bar{q}_{i-K}, \ldots, \bar{q}_i$ are now to be re-optimized jointly to yield the final ego pose estimate. Due to the absence of any additional external hardware like odometers or the like we resort to forcing the motion induced by these past one-shot estimates to follow certain dynamics. At this point we exploit the knowledge that the camera is mounted inside a vehicle which naturally follows non-holonomic motion models.

Thus, we augment each pose of the window by velocities in each dimension and set $r_i = (q_i, v_i)$ with velocity vector $v_i = (v_i^x, v_i^y, v_i^z)^T$. We refer to r_i as *velocity augmented pose*. Moreover, the time lag Δt_i between any two poses q_{i-1} and q_i is known. This allows to compute a prediction $\bar{r}_i$ of the velocity augmented pose r_i by applying the change $\Delta t_i \cdot v_{i-1}$ to pose q_{i-1}. A prediction of the velocity augmented pose r_i is thereby obtained from the prediction function

$$f(r_{i-1}, \Delta t_i) = f\left((q_{i-1}, v_{i-1}), \Delta t_i\right) \tag{5.7}$$

$$= \left(q_{i-i} \oplus \left(\Delta t_i \begin{pmatrix} v_{i-1} \\ \mathbf{0}_{3\times 1} \end{pmatrix}\right), v_{i-1}\right) \tag{5.8}$$

$$= \bar{r}_i \tag{5.9}$$

hence assuming a constant velocity vector. The prediction function is shown in Figure 5.3.

The pose adjustment then tries to balance the prediction $f(r_{i-1}, \Delta t_i)$ with its one-shot prior $\bar{q}_i$ while penalizing velocity changes. To this end, we define the subtraction of the velocity augmented poses

$$r' \ominus r = (q', v') \ominus (q, v) \tag{5.10}$$

$$= \left((q' \ominus q)^T, (v' - v)^T\right)^T \in \mathbb{R}^9 \tag{5.11}$$

and derive the error function

$$E_{\mathrm{adj}}(r_{i-K}, \ldots, r_i) = \sum_{k=0}^{K} ||q_{i-k} \ominus \bar{q}_{i-k}||^2_{\Sigma_{i-k}} +$$

$$\sum_{k=0}^{K-1} ||f(r_{i-k-1}, \Delta t_{i-k}) \ominus r_{i-k}||^2 \tag{5.12}$$

whose minimizing argument

$$\hat{r}_{i-K}, \ldots, \hat{r}_i = \underset{r_{i-K}, \ldots, r_i}{\arg\min} \left\{E_{\mathrm{adj}}(r_{i-K}, \ldots, r_i)\right\} \tag{5.13}$$

is taken as the final metric pose estimate.

The factor graph of (5.12) is depicted in Figure 5.4. The velocity augmented poses $r_{i-K}, \ldots, r_i$ which are subjected to optimization are shown. These are the arguments of the error function of (5.12). These are inter connected and each edge corresponds to one constraint. Edges connecting consecutive poses are constraints stemming from the motion model (prediction function (5.7)), are labeled with the time lag Δt and correspond exactly to the second summation of (5.12). Edges that connect to a one-shot prior $\bar{q}_i$ penalize any deviation of r_i to $\bar{q}_i$ and these edges represent the first summation of (5.12).

A joint re-optimization of a set of previous one-shot estimates increases the accuracy of the estimate. Additional constraints (motion model) provide additional

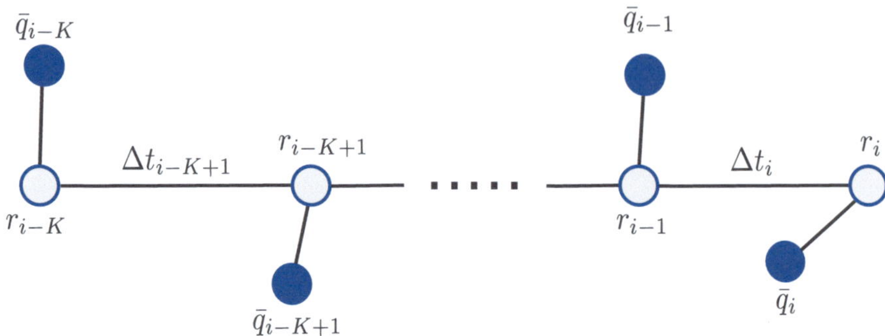

Figure 5.4: The factor graph associated with (5.12) is shown. The one-shot estimates $\bar{q}_i$ serve as a prior during pose adjustment. The priors are balanced with their predictions from the velocity augmented poses.

cues which can only be exploited in joint optimization. Furthermore, the squared Mahalanobis norm of the residual $\|\hat{q}_i \ominus \bar{q}_i\|_{\Sigma_i} = (\hat{q}_i \ominus \bar{q}_i)^T \Sigma_i^{-1} (\hat{q}_i \ominus \bar{q}_i)$ is interpreted to disclose any one-shot estimate $\bar{q}_i$ that is an outlier in the pose adjustment sense. If for any reason one such one-shot estimate has yielded an unreasonable value it is found hereby, pruned from (5.12) and the final estimate (5.13) is recomputed. This one-shot outlier detection further contributes to the overall robustness of the method.

Finally, we justify our choice of the constant velocity model to constrain the camera motion. Many more elaborate motion models like the curve linear models of [60] or dynamic single track models may appear deceptively tempting. However, these models require the knowledge of the mounting position of the camera relative to the vehicle center. The aforementioned models require a notion of the heading of the vehicle; not the heading of the camera which can be arbitrary in our case. Our goal was to allow this localizer to work without any troublesome camera to vehicle calibration. Nothing keeps one from using this approach with a sidewards facing camera even though we admit to have tested it only with forward and backward facing configurations. A possible extension to exploit motion models that are specifically tailored to car-like vehicles might be to integrate the camera-to-vehicle coordinate transform into the optimization objective and optimize everything jointly. We discuss this idea in more detail in the conclusion Chapter 7.

At this point we mention an alternative pose adjustment if a motion estimate between consecutive time steps is available. This motion estimate can be computed e.g. from vision itself or some external hardware may provide velocity and yaw rate estimates. Let the motion between poses p_{i-1} and p_i be denoted by $m_i \in \mathbb{R}^6$

such that $p_i = p_{i-1} \oplus m_i + \epsilon_i$ with an unknown noise vector ϵ_i. Then the motion model can be replaced by the motion constraint and (5.12) is replaced by

$$E_{\mathrm{adj}}(p_{i-K}, \ldots, p_i) =$$
$$\sum_{k=0}^{K} ||p_{i-k} \ominus \bar{p}_{i-k}||^2_{\Sigma_{i-k}} + \sum_{k=0}^{K-1} ||(p_{i-k} \ominus p_{i-k-1}) - m_{i-k}||^2 \qquad (5.14)$$

where the velocity augmented poses are now replaced by the regular poses. The velocities are not estimated anymore. The minimizing argument is found slightly faster and if the motion estimates are very accurate (e.g. from a high precision IMU) then it offers additional robustness. Then, however, an accurate 6 DOF camera-to-IMU calibration is required. We mention this approach only for completeness as it may be beneficial in some situations. In all experiments we use our vision only method of applying a motion model constraint to the window of past one-shot estimates. No external hardware is required anywhere in the experiments we present next.

5.3   Experiments

Next we present experiments on real world data to assess and evaluate our localization method. We examine each step of the localization method first. Then a measure for the overall localization accuracy is derived and evaluated on the test set. The presented method derives a map relative localization estimate of a map that does not have any global reference. Hence, ground truth is impossible to acquire. Therefore, we have chosen to use two independent cameras mounted on the same vehicle and record data for both of them. Then two completely independent localization estimates can be computed for each of the cameras. Finally, these two estimates are compared for consistency which yields a measure of localization accuracy which we believe is in the range of the overall localization accuracy.

After the aforementioned quantitative set of experiments we present some qualitative evaluations. We manually enhance the visual map with infrastructural objects like pedestrian crossings, traffic lights, lane markings, curb stones etc. During online localization these objects are projected into the camera that is believed to be at the estimated position. The projection is then overlaid onto the current camera image yielding an augmented reality system for intelligent vehicles. We present several screen shots of our augmented reality (AR) system.

5.3.1 Quantitative Experiments

We have equipped a standard station wagon with two stereo camera setups. One stereo rig is facing forward whereas the other is facing backwards. Imagery is recorded and analyzed thereafter. Note that forward and backward facing cameras are never used jointly but are always evaluated independently. Hence we obtain one set of recordings for each stereo setup. Stereoscopy is required and used only for the creation of the map. A mere single monocular camera is used in all localization experiments. No additional sensors like GPS or the like are used anywhere in the experiments. We note that the forward facing camera setup has a slightly narrower field of view which seems to impact some of the experiment.

We have picked a 7km route through mostly urban and partially rural areas as representative testing ground. We have travelled this route on three different days each two weeks apart. The first traversal was used to create the visual map and we will refer to this test set as *MAP*. The two remaining recordings are used for the localization experiments we refer to them as *LOC1* and *LOC2* respectively. A visual map was created for both backward and forward facing cases of the mapping test set *MAP* as presented in Chapter 4. Map computation is not time critical for real time operation and takes roughly three hours for each set. Almost thirteen million landmarks are created (for each case). The storage size for these map sizes is roughly 5 gigabytes each in a compact binary format. It includes all poses, landmarks and their visual descriptors.

In the following we will present several localization experiments which are performed on each of the test sets. In particular we use the mapping trajectory for the localization experiments as well. The localization experiment for the mapping case obviously yields excellent results and shall serve as an upper (or lower) bound which cannot be exceeded by any other test set. Thus, we obtain six results for each test which are *MAP*, *LOC1* and *LOC2* for the forward and backward facing camera configuration each.

At first we determine the traveled distance before a topological localization is possible, hence until the nearest pose of the map is found. We have replayed each test set from 500 equidistantly placed starting positions and determined the distance until a topological localization is achieved.

We present results as two sided violin plots in Figure 5.5. One plot represents one test set each. The plots show vertical and smoothed histograms of the distance until topological localization is possible. The left halves (blue) of each plot represents the results for the backward facing case whereas the right portion (orange) shows findings for the forward facing setup. The median is marked as well.

We have removed the upper 5% quantile from the plots for better visibility. The median for topological localization is 3.1 meters for the mapping test set

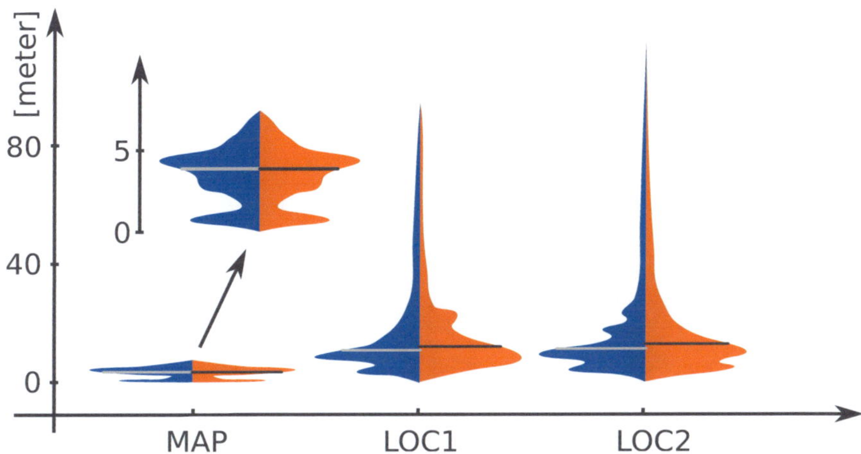

Figure 5.5: Two sided violin plots (vertical, smoothed histograms) for the required traveling distance until a topological localization is possible are shown for three test sets. The first plot shows results for the mapping sequence (shown again with more appropriate scale) whereas the others show findings for the two localization test sequences. The left half of each plot shows results for the backward facing configuration. The right portion represents the forward facing camera setup. The median is shown for each plot.

(both forward and backward), 8.0 meters for *LOC1/backward*, 9.1 meters for *LOC1/forward*, 7.8 meters for *LOC2/backward* and 9.6 meters for *LOC2/forward*. However, we note that some areas (especially rural ones) are unfavorable for topological localization and may easily require one hundred meters and more of traveling before topological localization is possible. Furthermore, our experiments indicate a high sensitivity to lane differences between mapping and localization. This can be seen from the spiking tops of the plots in Figure 5.5. All topological localizations that the localizer has output are corrected and are verified by the subsequent metric localization.

We follow the same testing procedure as before to assess the number of inlier landmark associations for each time step during one-shot estimation. To this end, each test set is used for metric localization and the number of inlier landmark associations (according to RANSAC) are tracked. Results are again shown by two sided violin plots for the six cases in Figure 5.6. Point matching works perfectly for the mapping trajectory since matching images are identical. Hence, the left plot of Figure 5.6 corresponds to the histogram of number of visible landmarks during mapping (every single landmark is associated correctly). Matching images from a

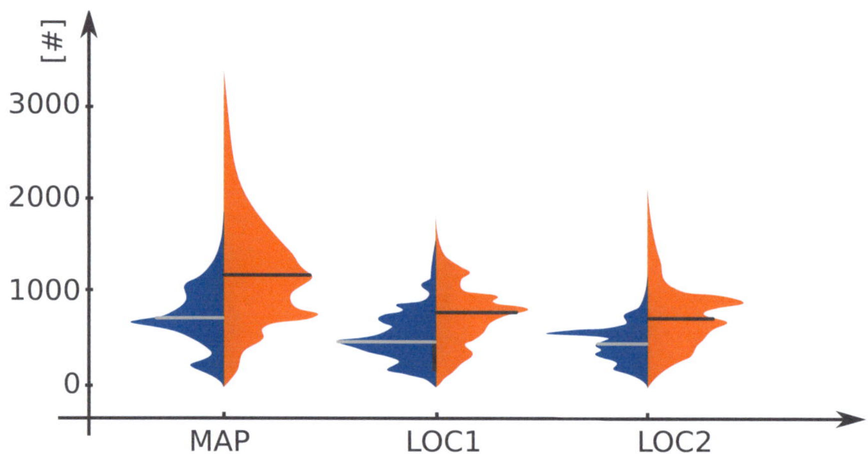

Figure 5.6: Violin plots for the number of successfully associated landmarks per camera image during localization are shown for the three test sets. See also Figure 5.5.

different day (*LOC1* and *LOC2*) is more realistic. The number of correctly found landmarks ranges from zero (underexposed camera in underbridge) to almost two thousand in some case. A significantly higher number of landmarks are associated for the forward facing case. We attribute this to the narrower field of view which makes camera calibration and feature matching easier.

Next, we illustrate the mean back projection error of the inlier landmark matches during one-shot estimation in pixels in Figure 5.7. Significant differences can be observed between forward and backward facing setups. We again attribute this to the narrower field of view which exhibits fewer distortions. The median back projection error is between 0.5 and 1.0 pixels; a range we would have expected for good localization.

So far only the left camera image of the stereo recordings has been used. Since we have recorded both left and right images of the stereo setup in all cases we are now able to estimate the trajectory for both the left and right camera. We compute the one-shot estimates for every left and right image independently and compare them for consistency. Since the base length of the stereo rig is known the right camera estimate can be compensated for it and subtracted from the left camera estimate. The norm of the difference between the two estimates are depicted in Figure 5.8. The left and right one-shot estimates clearly agree to within centimeter level accuracy. The experiment is repeated for the final localization after pose adjustment

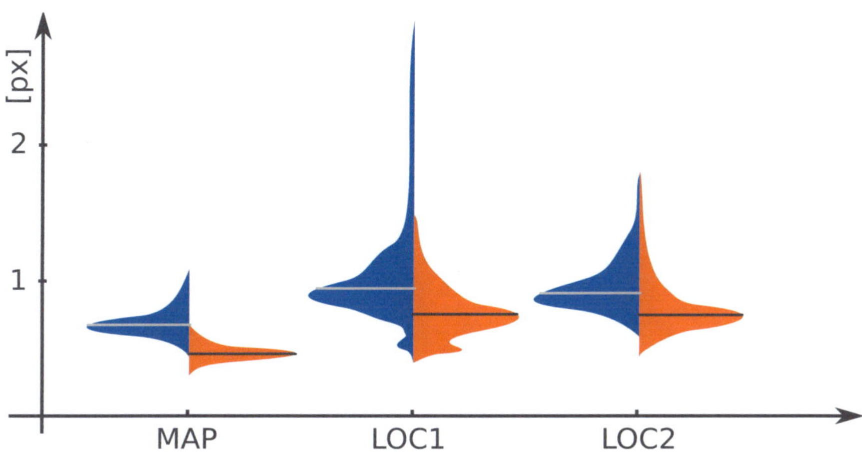

Figure 5.7: Violin plots for the mean back projection error of all inlier landmark associations (in pixels) are shown for the three test sets. See also Figure 5.5.

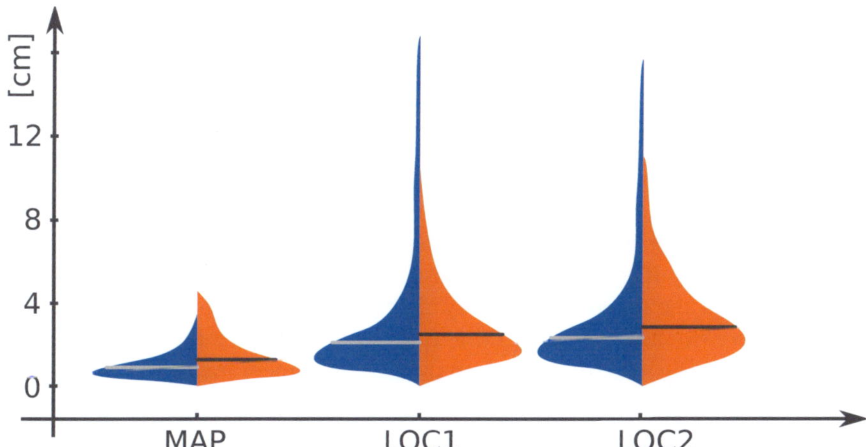

Figure 5.8: Violin plots of the difference of two independent localization estimates of two independent cameras mounted on the same vehicle are shown. They indicate the magnitude of the localization error relative to the visual map. Findings for the one-shot estimate are shown.

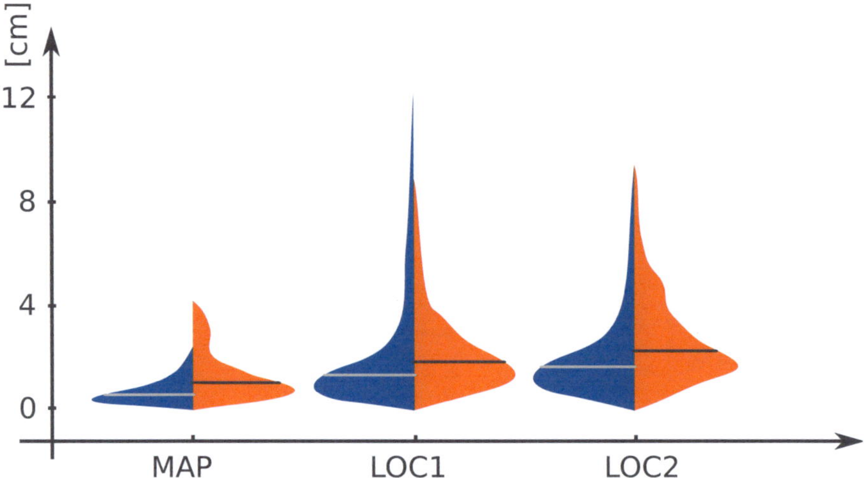

Figure 5.9: Violin plots of the difference of two independent localization estimates of two independent cameras mounted on the same vehicle are shown. They indicate the magnitude of the localization error relative to the visual map. Findings for the final ego pose estimate (pose adjustment) are shown.

and the results are visualized in Figure 5.9. We believe that the consistency measure presented here is of the same magnitude as the localization accuracy is.

5.3.2   Qualitative Experiments

Quantitative experimental findings were highlighted in the previous sections. Next, we present results of some qualitative tests. We have labeled objects of interest in some frames of the mapping trajectory. These can then be stored in map coordinates by reconstructing its 3D position from stereoscopy. During online localization these objects can be overlaid onto the current camera image. Thereto the 3D objects are simply projected into the camera which is believed to be at the estimated pose.

  The labeling is depicted in Figures 5.10 and Figure 5.11. The object of interest (e.g. pedestrian crossing) is manually labeled in one image and stereoscopy is used to compute its 3D camera relative position. Since the pose of the camera is already known from the mapping process the camera relative position can be translated into a global (map) relative position and stored in the map.

Figure 5.10: The labeling of a pedestrian crossing is shown. Line segments drawn on top of one image are used with stereoscopy to derive its 3D map relative position from it. This information is stored in the map data structure.

During localization a high precision 6 DOF ego pose estimate is available. This allows to load all infrastructural objects of the vicinity and project it into the current camera image. One such example is shown in Figure 5.12. It is shown how the previously labeled pedestrian crossing is approached. At first it is even occluded by the truck but its camera relative position is nevertheless very accurate. Moreover, this approach allows to make these objects available long before any sensor would be able to detect it. Its "sensing" range is unbound. To demonstrate this effect we show the same scene in Figure 5.13 but a few ten seconds earlier. The crossing is still several hundred meters away but accurately overlaid.

Figure 5.14 again shows the same area but these images were recorded two weeks later. Hence, they are from a different test set. The road surface is wet in Figure 5.12 whereas it is dry in Figure 5.14. However, the exact same visual map was used. The AR system works well and with the required precision.

Yet another example is illustrated in Figure 5.15. The examples show a good fit of the project objects with the camera image. A false localization would be seen be a significant deviation of the objects projection from the actual image position. Moreover, a six DOF estimate is necessary for such systems. Pitch and roll needs to be compensated for.

Figure 5.11: Any object of interest can be labeled. Here a tunnel entrance and a traffic light is marked.

Finally, we cannot resist to mention that the system was used for autonomous driving with the IMU extension mentioned in Section 5.2. We were able to successfully drive fully automatically in a dense urban environment using this localization alone while fully neglecting any GPS readings from the on board receiver.

Figure 5.12: Results of our AR system are shown. The previously labeled pedestrian crossing is reliably made available during online localization. This approach allows to circumvent any problems caused by occlusion.

Figure 5.13: The system allows to make the map objects available long before any sensor would be able to detect it. The pedestrian crossing of Figure 5.12 is shown some ten seconds earlier.

Figure 5.14: The same scene as before is shown but with a recording of a different day.

Figure 5.15: Another screen shot from the AR system. Traffic lights and a tunnel entrance are exemplary shown.

6   Descriptor Learning

Many problems in computer vision and pattern recognition rely on the ability to uniquely describe the neighborhood of a single pixel of an image by one numeric feature vector. To this end, the pixel neighborhood around the pixel in question is analyzed and transformed into a compact vector representation. This vector representation can thereafter be used to address a variety of different vision problems. Finding pixel positions in a set of images that belong to the same point in 3D is a prominent and frequent example. Furthermore, a collection of such feature vectors can be used to describe the content of an image which in turn allows to query large databases in order to find the same image content from an appearance based perspective. Alternatively, these feature vectors can be used to assign one class label of a set of known classes to an image. Image classification helps to automate many tasks that are otherwise tedious and error prone if performed by humans.

Throughout the rest of this chapter one algorithm that extracts a vector from a pixel neighborhood is referred to as a *descriptor* whereas the vector itself is referred to as *feature vector* or feature for short.

The literature offers a plethora of different descriptors that have recently been published and the flood of newly published methods doesn't seem to cease. Many descriptors are now accepted as standard methods in this area such as SURF [6], BRIEF [13], SIFT [48], HOG [18] and more as they perform well on various tasks. However, all these methods are general purpose descriptors designed to perform well on average. One might imagine that a problem specific descriptor may outperform a general purpose one on that specific problem.

In this chapter we present a set of elementary algorithmic building blocks that can be combined in any order to form a descriptor for each such combination. We show how many standard descriptors can be decomposed into these blocks. Hence, a systematic construction of descriptors is thereby possible. Furthermore, we argue that these blocks allow to automatically create novel descriptors by random combination. A preliminary sketch of the concept has been previously presented in [36].

In the following we present a method to combine these building blocks to automatically create novel descriptors. Then, a fitness function that evaluates one such descriptor for a specific problem is proposed. The fitness function is used to assess each one of the descriptors and a genetic algorithm selects the best performing ones. The best performing ones are slightly modified by a mutation operator to create a new generation of offspring descriptors which replaces the parent population. By iterating this procedure, we automatically learn a descriptor that is tailored to the problem of interest. In our example, we address the problem of visual place recognition under varying illumination conditions. We show the feasibility of the approach by comparing the automatically learned descriptor to state-of-the-art methods and demonstrate a substantial improvement.

First, we present the building blocks used to span the space of all descriptors in Section 6.1. Thereafter, evolution strategies which we use as a genetic algorithm are elucidated in Section 6.2. Experiments and the results of the descriptor learning are presented in Section 6.3.

6.1 Algorithmic Building Blocks

We propose to use elementary building blocks which can be chained in any order to create many descriptors. First, we present a simple example and detail the specific building blocks thereafter.

We assume a pixel position u, v is given (e.g. by some detector) and we strive at extracting an image feature vector f describing the neighborhood of that pixel position. The gray scale image I can be convolved with a high pass and a low pass filter to yield the filtered responses $I_{\mathrm{low}} = I * h_{\mathrm{low}}$ and $I_{\mathrm{high}} = I * h_{\mathrm{high}}$ with $*$ denoting convolution and h being the filter kernels. A simple descriptor f_1 can then simply take the responses at the pixel position as a 2-vector output $f_1(u, v) = (I_{\mathrm{low}}(u, v), I_{\mathrm{high}}(u, v))^T$. Such a descriptor can be extended by an additional processing step which converts Cartesian to polar coordinates by $f_2(u, v) = (\angle f_1(u, v), \|f_1(u, v)\|)$. Note that such an operation does not require knowledge of how f_1 was computed. Still, such descriptor is low dimensional and not very discriminative. To extend the descriptor f_2 it can be applied several times at predefined offsets around the pixel position u, v and use the concatenation as output. Hence, let $(u_1, v_1), \ldots, (u_N, v_N)$ be pixel offsets. Then f_2 can be extended to yield the descriptor $f_3(u, v) = (f_2(u + u_1, v + v_1), \ldots, f_2(u + u_N, v + v_N))$ by applying it multiple times. This operation is again independent of the previous processing step. It is always possible to apply any given descriptor multiple times without having to know how that descriptor was computed. This chain of operations can be extended almost arbitrarily. Another possible extension is to compute f_3 in a circular neighborhood of u, v and compute a histogram from these f_3-values. The histogram itself is then taken as f_4.

This example descriptor can be summarized by its processing steps: low/high pass filtering, Cartesian to polar transform, repetition and finally histogram computation. The point is, that each step can be replaced by any alternative. The low/high pass filters could be replaced by derivative filters. The polar transform could be skipped or replaced by a vector normalization operation or any other sensible operation. The repetition could change its offset pixel position or the histogram bin centers (or number) could change. Moreover, steps can be repeated several times. Nothing keeps one from applying yet another repetition at the end to yield yet another descriptor f_5. We note that the descriptor presented above serves as a mere toy example to introduce the basic idea. It is not meant to be functional as an image descriptor.

In the next paragraphs, each of the possible building blocks is introduced in greater detail. Every descriptor starts with some filter operation and is followed by any number of additional steps. For most of the elementary processing steps there exists a set of variants of that step. The repetition pattern for instance can have many different sets of offset pixel positoins. These variants are explained as well.

Filter Banks: The following filter banks are contained in our scheme:

- Sobel Filters which create a two-vector output. This filter bank does not have alternative variants. The two filter kernels are given by

$$F_{\text{Sobel}-\text{hor}} = \begin{pmatrix} 1 & 2 & 1 \\ 0 & 0 & 0 \\ -1 & -2 & -1 \end{pmatrix} \tag{6.1}$$

$$F_{\text{Sobel}-\text{vert}} = \begin{pmatrix} 1 & 0 & -1 \\ 2 & 0 & -2 \\ 1 & 0 & -1 \end{pmatrix} \tag{6.2}$$

  and detect edge like structures in images [32].

- Gaussian blurring which convolves the input image with a truncated Gaussian kernel and thereby creates a one-vector for each pixel position. We provide a set of kernel widths σ to control the degree of smoothing as variants with $\sigma \in \{3,5,7,9,11,13\}$.

- Derivative in horizontal and vertical direction which is loosely related to the

Sobel filter bank above but without smoothing. We apply a difference filter

$$F_{\text{diff}-\text{hor}} = (-1,\underbrace{0,\ldots,0}_{=Z},1) \tag{6.3}$$

$$F_{\text{diff}-\text{ver}} = \begin{pmatrix} -1 \\ 0 \\ \vdots \\ 0 \\ 1 \end{pmatrix} \tag{6.4}$$

in horizontal and vertical direction. The step width Z denotes the number of zeros and are provided as the variants of this filter bank. Its range is $Z \in \{0,\ldots,4\}$. The filter response is two dimensional.

- A Haar filter bank is provided which applies three filters at each scale defined by

$$F_{\text{haar}-\text{vert}} = \frac{1}{2^{2s-1}} \begin{pmatrix} 0 & \cdots & 0 & 1 & \cdots & 1 \\ \vdots & \ddots & \vdots & \vdots & \ddots & \vdots \\ 0 & \cdots & 0 & 1 & \cdots & 1 \end{pmatrix} \in \mathbb{R}^{2^s \times 2^s} \tag{6.5}$$
$$\underbrace{\qquad\qquad}_{2^{s-1}}$$

$$F_{\text{haar}-\text{hor}} = \frac{1}{2^{2s-1}} \begin{pmatrix} 1 & \ldots & 1 \\ \vdots & \ddots & \vdots \\ 1 & \ldots & 1 \\ 0 & \ldots & 0 \\ \vdots & \ddots & \vdots \\ 0 & \ldots & 0 \end{pmatrix} \in \mathbb{R}^{2^s \times 2^s} \tag{6.6}$$

$$F_{\text{haar}-\text{diag}} = \frac{1}{2^{2s-1}} \begin{pmatrix} 1 & \ldots & 1 & 0 & \ldots & 0 \\ \vdots & \ddots & \vdots & \vdots & \ddots & \vdots \\ 1 & \ldots & 1 & 0 & \ldots & 0 \\ 0 & \ldots & 0 & 1 & \ldots & 1 \\ \vdots & \ddots & \vdots & \vdots & \ddots & \vdots \\ 0 & \ldots & 0 & 1 & \ldots & 1 \end{pmatrix} \in \mathbb{R}^{2^s \times 2^s} \tag{6.7}$$

where s defines the scale. The filter bank consists of all scales $s \in \{1,\ldots,N\}$ and $N \in \mathbb{N}$ defines the variant. We provide several variants

$N \in \{1,\ldots,4\}$. A Haar filter bank of maximum depth N contains $3N$ filters.

- The rank transform is provided as a filter bank that is not a convolution operation. Let $I(u,v)$ be the gray level of pixel position u,v. Then the rank transform outputs

$$R(u,v) = \sum_{k=1}^{8} \mathbb{I}(I(u,v) > I(u_k,v_k)) \tag{6.8}$$

with $(u_1,v_1),\ldots,(u_8,v_8)$ being the eight immediate neighbor pixel positions of u,v. Hence (6.8) computes the number of neighbor pixels that are brighter than $I(u,v)$ where $\mathbb{I}(b)$ is the indicator function which evaluates to one if the Boolean expression b is true and to zero otherwise. The is only one variant of the rank transform.

- Similarly to the rank transform the census transform is provided as a filter. The census transform for pixel position u,v is defined by

$$C(u,v) = \sum_{k=1}^{8} \mathbb{I}(I(u,v) > I(u_k,v_k)) \cdot 2^{(k-1)} \tag{6.9}$$

with (u_k,v_k) again being the neighbor pixel positions to u,v as above. In contrast to the census transform, the difference of gray values is first translated into a binary representation $(\mathbb{I}(I(u,v) > I(u_1,v_1)),\ldots,\mathbb{I}(I(u,v) > I(u_8,v_8)))$ and then interpreted as a decimal number. The census transform only contains one variant.

- Last, we added the identity filter to the set of filter banks which leaves the input image untouched.

Repetition: As in the example above, our scheme requires the ability to repeat a given descriptor several times. The repetition block evaluates a given descriptor at a set of predefined offsets around u,v. The features obtained from these offset positions are concatenated and form the output of this block. Every set of offsets is one variant that can be chosen for this block.

Formally, we define a set of pixel offsets

$$\mathcal{O}_N = \{(u_1,v_1),\ldots,(u_N,v_N)|(u_n,v_n) \in \mathbb{Z} \times \mathbb{Z}\} \tag{6.10}$$

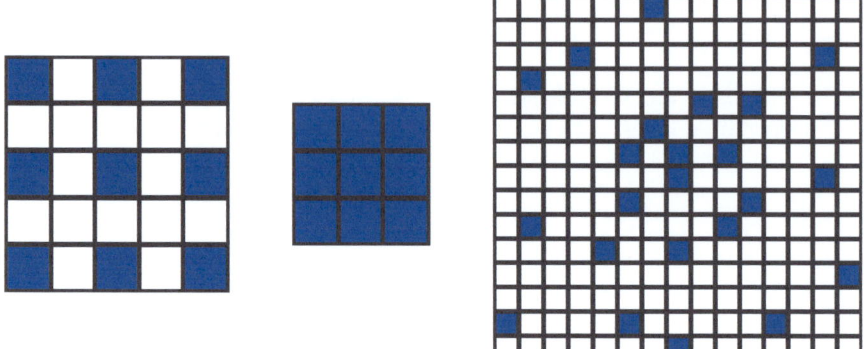

Figure 6.1: A given descriptor can be extended by applying it at different offsets of pixel positions and concatenation. Several such repetition masks are shown.

which are integer values. A given descriptor A is extended to yield

$$B(u, v) = \begin{pmatrix} A(u + u_1, v + v_1) \\ \vdots \\ A(u + u_N, v + v_N) \end{pmatrix} \tag{6.11}$$

by applying it N times around the pixel position u, v and concatenating the respective outputs. We chose three different rules to generate an offset index $\mathcal{O}_N$.

- The first set of offsets is a repetition pattern quite close to u, v. The number of offsets for these patterns are between nine and sixteen.

- Moreover, a set of BRIEF [13] like repetition patterns for different numbers of offsets are stored. These offsets follow Gaussian distributions hence exhibiting a closer density towards the center. The number of offsets N ranges from 20 to 300 for these patterns.

- Finally, a small set of repetitions are stored as variants which are further away from the center and form an equidistantly spaced grid. The number of offsets N is between four and sixteen. This repetition is commonly found as the last step in descriptors like SURF [6].

Some of the first two sets of repetition patterns are exemplary shown in Figure 6.1.
Linear Transform: A matrix multiplication may be appended to a sequence of construction steps. If a descriptor computes the feature vector A for pixel position

$$
\begin{bmatrix}
0 & 0 & 1 & 0 & 0 & 0 & 0 & -1 & \cdots \\
0 & 0 & 0 & 1 & -1 & 0 & 0 & 0 & \cdots \\
0 & -1 & 0 & 0 & 0 & 0 & 1 & 0 & \cdots \\
1 & 0 & 0 & 0 & 0 & -1 & 0 & 0 & \cdots \\
& & & \vdots & & & & \vdots &
\end{bmatrix}
\begin{bmatrix}
-1 & 0 & 0 & 0 & 1 & 0 & 0 & 0 & 0 \\
0 & -1 & 0 & 0 & 1 & 0 & 0 & 0 & 0 \\
0 & 0 & -1 & 0 & 1 & 0 & 0 & 0 & 0 \\
0 & 0 & 0 & -1 & 1 & 0 & 0 & 0 & 0 \\
0 & 0 & 0 & 0 & 1 & -1 & 0 & 0 & 0 \\
& & & \vdots & & & & \vdots &
\end{bmatrix}
$$

Figure 6.2: A given descriptor can be extended by applying a constant matrix M to it. Two example matrices are shown. See text for details.

u, v then this can be altered by multiplying $B = MA$ for a constant matrix M. Two sets of predefined matrices are stored. Let N be the dimension of the input feature A .

- One set contains $2N \times N$ matrices with exactly one -1 and one 1 in each row such that every column contains exactly one non-zero entry.

- The second set contains $N-1 \times N$ matrices with one column of 1s and rows containing exactly one additional -1 such that every column has at most one -1.

One example each is shown in Figure 6.2. Each of this configurations serves as one variant.

Non-linear Transform: Some descriptors compute a polar coordinate representation from e.g. horizontal and vertical derivatives such as e.g. HOG and SIFT. Such a transform is obviously non-linear and we provide three types of such non-linear transforms. They form the set of variants for this building block.

- One Transform that computes polar from Cartesian coordinates is provided. For a feature $A = (a_1, \ldots, a_N)^T$ as input it outputs $B = (\sqrt{a_1^2 + a_2^2}, \mathrm{atan2}(a_1, a_2), \ldots, \sqrt{a_{N-1}^2 + a_N^2}, \mathrm{atan2}(a_{N-1}, a_N))^T$. For odd N the last a_N is simply taken as is.

- Another nonlinear transform that has proved to be useful is the transform that scales a given feature to unit norm. That is it outputs $B = \frac{A}{||A||_2}$ for input feature A.

- Finally a nonlinear transform inspired by a recent trend of using binary descriptors [30] is added to the set of variants. A sign quantization transform which replaces all negative elements of a vector by zeros and nonnegative elements by ones is provided.

Histogram: The histogram building block is one of the most powerful blocks to extend a given descriptor. If a descriptor B shall describe the neighborhood of the pixel position u, v and some descriptor A is already available then B can apply A N times in a predefined neighborhood $\mathcal{O}_N$ around u, v (e.g. a circular area.), collect all features vectors $a_1, \ldots, a_N$ of this neighborhood and compute a histogram from this set of features. The histogram itself is then the feature that B computes.

We chose a very general and broadly applicable histogram representation. A histogram is a set of histogram bin centers and the number of features that are assigned to each bin. Bin assignment is computed by nearest neighbor search. The bin centers, however, are constant and thus the histogram is completely represented by their bin counts. Note that this general definition equally represents simple one dimensional histograms as well as histograms of any high dimensional space.

Variants of this block are two-tuples $(\mathcal{C}, \mathcal{O}_N)$ where $\mathcal{C} = (c_1, \ldots, c_N) \subset \mathbb{R}^D$ is an ordered set of N bin centers with D being the dimension of the input descriptor A that shall be extended. The area at which the input descriptor is evaluated is defined by the neighborhood $\mathcal{O}_N \subset \mathbb{Z}^2$ of pixel position offsets (cf. Repetition). Hence, the input dimension of this block is D and the output feature is of dimension N as there are N bin centers c_n.

It remains to show how to compute sensible bin centers for a given input descriptor A. We therefore provide a set of sample images and pixel positions and the bin centers are drawn from evaluating A at N pixel positions of these sample images. Thereafter the bin centers are kept constant. The process is depicted in Figure 6.3.

Dimensionality Reduction: As the dimension of a descriptor which is a random combination of these building blocks may easily grow very large it is necessary to provide a dimensionality reduction block. The dimension of a feature vector may be reduced by multiplying the vector by a constant matrix M. For dimension reduction we use random projection matrices ([10]). The degree by which the vector is reduced is the set of variants for this step. It ranges from 10% to 90%.

Summation: A summation block is presented next. Let a given descriptor A compute the features $a_1, \ldots, a_N$ in some neighborhood $\mathcal{O}_N$ around the pixel position u, v. This first step is much like the first step of the histogram computation. Then the output can be the summation of these vectors e.g. $B = \sum_{i=1}^{N} a_i$ where the summation extends over all pixels of the neighborhood $\mathcal{O}_N$. Alternatively the absolute value can be summed by $B = \sum_{i=1}^{N} \mathrm{abs}(a_i)$ where the absolute function is applied component-wise. These two options are varints of this block. A third

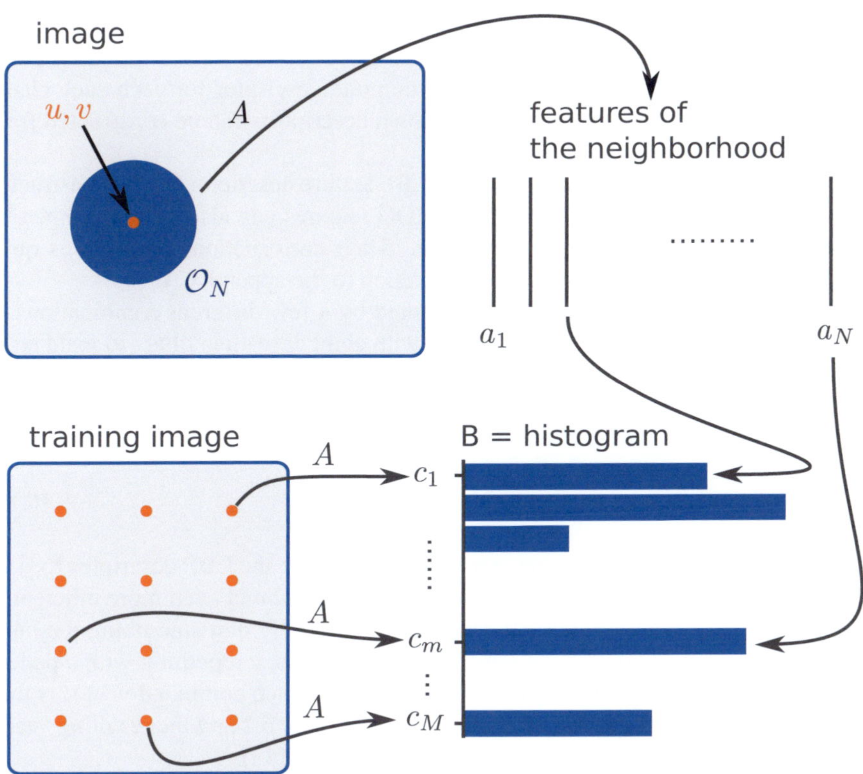

Figure 6.3: The pixel position u, v shall be described and a descriptor A is available. Then, A is evaluated on all pixel positions of the neighborhood $\mathcal{O}_N$ forming the features $a_1, \ldots, a_N$. These are then assigned to the nearest neighbor bin center $c_1, \ldots, c_M$ of the histogram. These bin centers are constant and never changed. They are initially computed by evaluating A on predefined pixel positions of an independent training image. The histogram is then the output of the new (extended) descriptor B.

variant is to concatenate both versions e.g.

$$B = \begin{pmatrix} \sum_{i=1}^{N} a_i \\ \sum_{i=1}^{N} \mathrm{abs}(a_i) \end{pmatrix}. \tag{6.12}$$

Choosing a set of such building blocks and selecting one variant for each block now allows to chain these blocks and obtain one descriptor for each such chain. Next, we demonstrate how commonly known descriptors can be constructed from our set of building blocks.

We exemplary show how the BRIEF and LBP feature descriptor can be constructed using the proposed blocks. SURF and HOG features can also be represented by a sequence of blocks as proposed above. Their construction, however, is quite technical and we therefore move the derivation to the appendix A.

The LBP feature can actually be constructed by a few different combinations of steps. The input image can be convolved with eight derivative filters to yield an 8-vector for the pixel position u, v. This vector is thereafter sign quantized to yield a 1 for every nonnegative entry and 0s otherwise. A linear transform is then applied with the following matrix

$$M = \begin{pmatrix} 2^0 & 2^1 & \dots & 2^7 \end{pmatrix}. \tag{6.13}$$

Computing a histogram of these values finally leads to the LBP descriptor [53]. BRIEF is widely used as it is very efficient to compute and even more efficient to match due to its binary nature. It can be constructed by first smoothing the input image with a Gaussian mask of appropriate size. Then, a repetition with a pattern as depicted on the right of Figure 6.1 is applied. A such computed vector is then multiplied by a matrix as depicted on the left of Figure 6.2 and the resulting vector is sign quantized. This yield exactly the descriptor of [13].

6.2   Learning Procedure

We are interested in finding a descriptor which is specifically tailored to the problem at hand. As mentioned in the introduction of this chapter, image descriptors find their application in a broad field of different domains. A descriptor performing well in one domain may not necessarily perform equally well in another. To quantify the performance of a given descriptor for a specific domain we introduce the notion of a fitness function. A fitness function takes a descriptor (hence the entire algorithm) as input and outputs a value that describes the performance of this descriptor for one specific problem. If for instance feature matching between

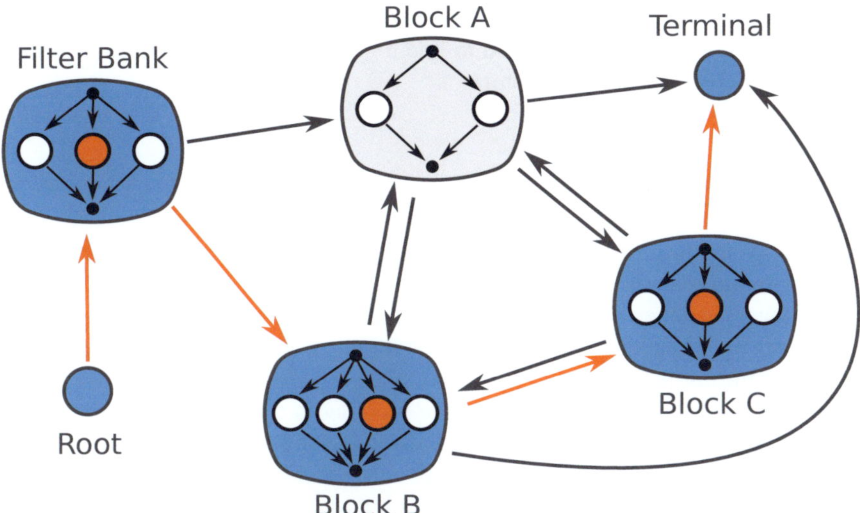

Figure 6.4: A simplified example graph of processing steps. Large nodes (blue/gray) are processing steps and small nodes inside (orange) are variants. A path through the graph corresponds to exactly one descriptor. A path is indicated by the orrange arrows and all nodes of the path are blue. It always starts with one variant of a filter bank and proceeds through additional processing steps (e.g. repetition, histograms, summation etc.).

images is the domain of interest one could evaluate a set of feature matches between these images against ground truth matches and take the number of correctly associated points as the fitness. Herein, we seek a descriptor which is well suited for place recognition under varying illumination conditions. The associated fitness function is introduced in Section 6.2.1

The descriptor construction scheme as presented above can be summarized by a graph. The nodes of the graph represent the building blocks. Each node contains a constant set of variants. Every path through the graph corresponds to exactly one descriptor. The graph is depicted in Figure 6.4. Using this graph/path notation it is easy to randomly create a large set of descriptors by sampling paths through the graph. We have constrained the maximum length of a path (number of steps) to six. But even then the total number of different descriptors which can be constructed reaches almost a billion due to the combinatorial explosion.

Hence it is difficult to find the best performing descriptor for a particular computer vision problem. In this section we show how we search for a good performing descriptor. We introduce the fitness function that evaluates any given descriptor for suitability for the specific problem of place recognition. Thereafter, a meta heuristic is reviewed which is used to find a good performing descriptor from the huge set of constructible descriptors.

6.2.1 Fitness Function

We follow the line of Chapter 4 and describe one image of a sequence by one holistic feature vector. The image is down sampled and tiled into 4×4 equally sized square patches of small size. The center pixel position of every patch is described by a feature vector and these single tile features are concatenated to represent the entire image.

To evaluate a given descriptor we have recorded a training sequence. We have traveled a route of approximately 10km on three different times of the day. We recorded image sequences and established correspondences of images showing the same place. The image pairing was computed automatically from high precision GPS measurements. We refer to these image pairs as positive pairs. Furthermore, we randomly created negative pairs of images that do not show the same place. Example images are shown in Figure 6.5. The recordings are from a day with clear sky and harsh cast shadows are clearly visible. The first three images of Figure 6.5 show the exact same place. The overall appearance of that place varies considerably.

Using this training set as ground truth, a precision-recall (PR) curve can be computed by varying a classification threshold of norms of feature vector differences. That means we strive to find a descriptor that best separates the positive from the negative pairs. By including images recorded in the morning, at noon and evening the change that naturally occurs during the course of a day is well captured in our training set.

6.2.2 Evolution Strategies

The fitness function evaluates any given descriptor and thereby allows to compare descriptors. Next we present the search strategy which searches for the best performing one. The search does not guarantee to find the single best performing combination of processing steps but still seeks a combination which is "locally" optimal. Common optimization strategies like Gauss-Newton or Powells method [57] cannot be applied since the set of descriptors cannot be modeled as a vector space of fixed dimension. Thus, we resort to evolution strategies [9].

Figure 6.5: Some sample images from the training set. The first three show the same place during morning, noon and evening. Note the severe changes in illumination and cast shadows.

At first, a set of M descriptors are randomly created by sampling paths through the graph of processing steps. Each of these M combinations is then evaluated by the fitness function and the best $L < M$ performing ones are kept whereas the rest is discarded. These L descriptors are slightly modified by a mutation operation to yield another set of M child descriptors. These are again evaluated and only the top scoring ones are selected into the next generation. Mutation and selection are iterated until convergence.

It remains to show how a given path through the graph of processing steps can be mutated. We implemented three different mutation operations.

- A path can be extended by a single additional processing step. This processing step does not necessarily need to be added to the end of the path but can also be added in between. To keep the computational complexity of the resulting descriptor tractable we have bounded the maximal number of steps to six. This mutation operation is hence only possible if the number is not reached yet.

- The inverse of the previous step is also possible. A single processing step can be removed anywhere from the path.

- Finally, a variant of a processing step can be exchanged for another one.

Figure 6.6 shows three possible child descriptors of the one of Figure 6.4 that are created by each of the three possible mutations. To speed up the search process, we have thinned out the training set by removing easy to distinguish image pairs that do not contribute much to a training success as proposed in [7]. Image pairs that are correctly classified by many descriptors are considered for removal in favor of more challenging pairs. Thereby we are able to approximately double the number of descriptor evaluations per time.

6.3   Experiments

We ran the evolution strategy for the fitness function for roughly one hundred hours. We have tested several parameters for L and M with one to four parents and four to sixteen children. We observed that the performance of the current generation of descriptors improves drastically in the beginning but slows down thereafter. Better performing descriptors are successively harder to find once the learning process proceeds. At the point of convergence where no children improve

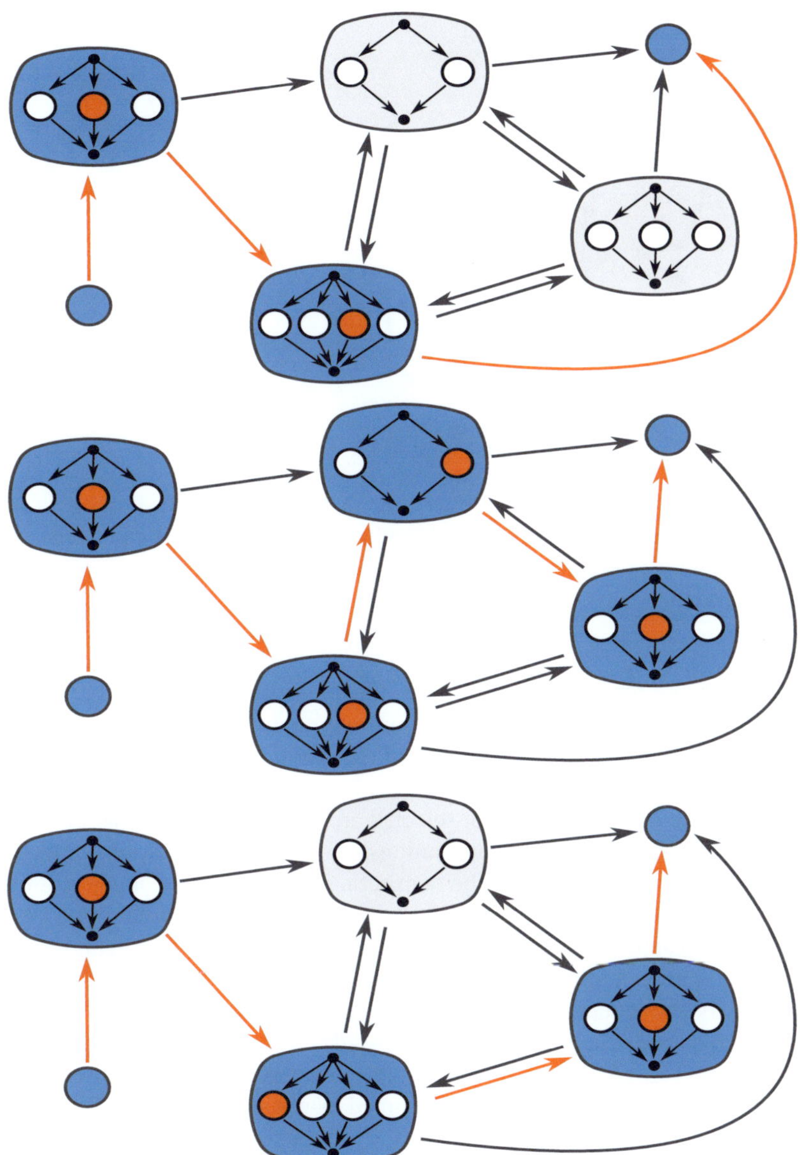

Figure 6.6: The path trough the graph of Figure 6.4 can be mutated into any of these paths in one mutation step.

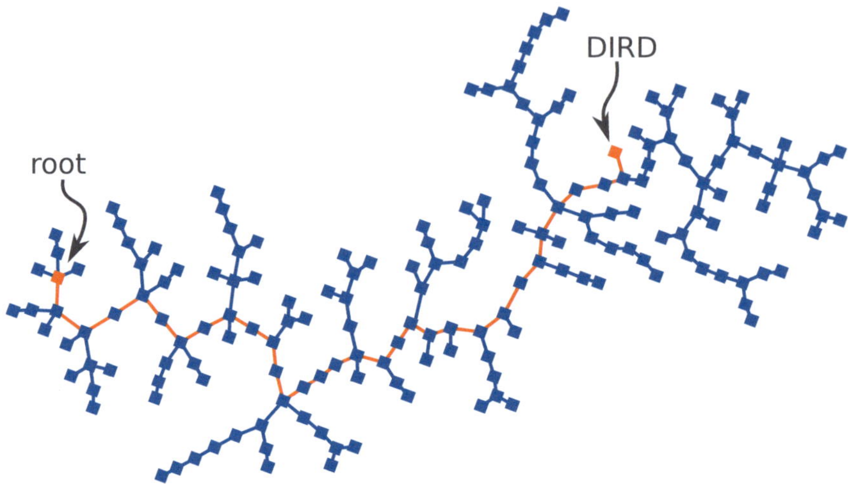

Figure 6.7: A "family tree" of descriptors during the learning process is shown. One node denotes one descriptor and the links denote a child-parent relationship. The root descriptor creates many descendants until the best performing descriptor is finally found.

the current score for several generations the top scoring element achieves an area under the PR curve (AUC) of 0.82 for our training dataset.

Figure 6.7 shows the entire evolution of the learning process. The graph shows one node for each descriptor that was tested and thereafter inherited into the next generation (the ones that were not selected are not shown). An edge between any two nodes denotes a child-parent relationship. Hence, the root descriptor shown on the left was mutated to form the connecting nodes etc. The orange path shows the path from the root descriptor to the best performing one on the right.

Unlike other automatically learned data structures like e.g. artificial neural networks our approach allows to interpret the learned descriptor as each processing step is easily understood. The automatically learned descriptor convolves the input image with a set of Haar filters on four cascades in vertical, horizontal and diagonal direction (cf. (6.5)-(6.7)) . Each vector corresponding to the filter responses is thereafter scaled to unit norm and summed over a set of sixteen offset positions. Finally, nine such vectors of an equidistantly spaced grid around the pixel in question are concatenated to form the final feature vector. The feature vector is 216 dimensional and we dub the descriptor DIRD (Dird is an Illumination Robust Descriptor).

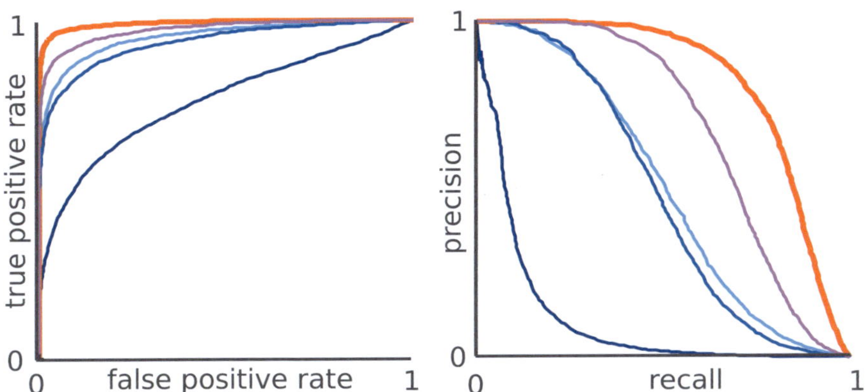

Figure 6.8: ROC and PR curves for the test set. From bad to good: simple gray value descriptor, SURF, BRIEF, U-SURF and the learned descriptor DIRD. See also Figure 6.9.

To assess the proposed learning approach we have tested DIRD on an independent test set and compared its performance to state-of-the-art alternatives. We again formed pairs of images showing the same place and pairs of images that do not show the same place. However, the image pairs are computed from recordings that are completely disjoint from the training set to avoid any biases towards DIRD. Computing the AUC on the test set we found that a simple gray value descriptor yields 0.11, SURF achieves 0.49, BRIEF scores 0.51 and U-SURF evaluates to 0.68. Finally, DIRD achieves an AUC of 0.82 substantially outperforming any of the aforementioned descriptors. The associated PR and ROC curves are shown in Figure 6.8 whereas the AUC plots are depicted in Figure 6.9.

 DIRD was learned with place recognition in mind and shows good performance in this setup as presented before. Next, we evaluate its performance in point matching tasks which require illumination robustness. Reliable point matching is required oftentimes in a variety of robotics applications. Visual odometry and localization (see Chapter 5) are some examples. Moreover, point matching can become computationally complex as many distance measures between feature vectors need to be computed. To be computationally tractable we propose two additional variants of DIRD. Generally each component of a DIRD feature vector is of type `float` which requires four bytes of data storage each. For fast feature matching, however, byte vectors are more convenient to compare (due to recent advances in CPU architectures offering SIMD instructions). Hence we propose to quantize the float values into 256 discrete byte values. A quantization can be represented by a set of intervals $\mathcal{Q} = \{[q_i, q_{i+1}) | i = \{0, \ldots, V - 1\}\}$ with $V \in \mathbb{N}$ being the number of

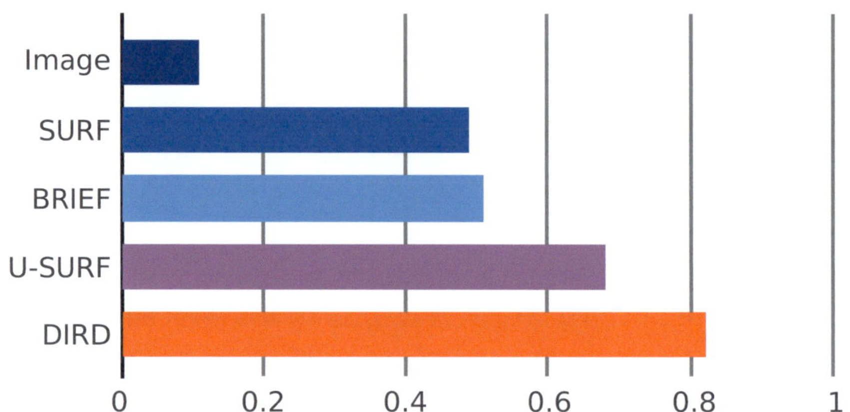

Figure 6.9: Area under the PR curve for a simple gray value descriptor (Image), SURF, BRIEF, U-SURF and the learned descriptor DIRD. For the curves see Figure 6.8.

quantization values. In this case, we consider $V = 256$ byte values. The quantization then maps any value $v \mapsto i$ if $v \in [q_i, q_{i+1})$.

Each component of the feature vector can then be quantized using such a quantization. Note, that it is easily possible to use different quantizations for different components of the feature vector. In fact, we use two different quantizations $\mathcal{Q}_1$ and $\mathcal{Q}_2$ for the odd and even components of a DIRD vector. Figure 6.10 shows the histograms of the odd and even components of a set of DIRD vectors. Their shape differs substantially which is why we handle them differently.

Examining the feature value distributions of Figure 6.10 reveals values of very high occurrence probability (e.g. a value of zero for the even components). We design our quantizations so that it has a finer resolution in these areas i.e. the interval boundaries q_i and q_{i+1} are much tighter. An easy way to achieve such a quantization is to compute V quantiles of the occurring values and take these as the interval boundaries q_i. This automatically yields a finer quantization in regions of high occurrence. Note, that the feature values are (approximately) uniformly distributed after such a quantization.

Following recent trends in computer vision to use binary descriptors [30] we use a third flavor of DIRD which quantizes each component to either 0 or 1. We again determine the quantization threshold by using the 50% quantile for the odd and even components of a set of DIRD features. A binary-quantized DIRD vector re-

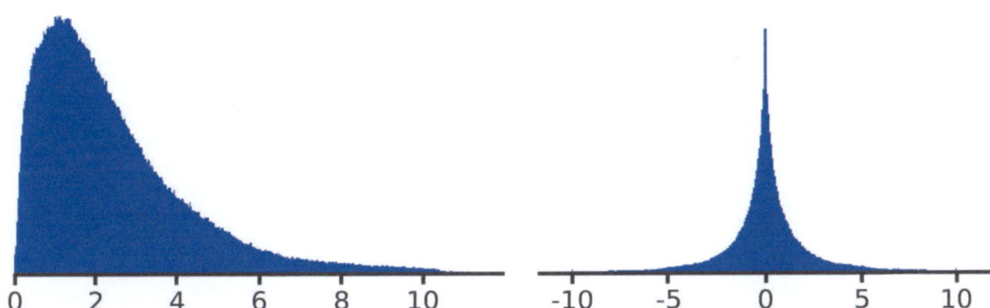

Figure 6.10: Histograms of values of the odd (left) and even (right) vector components of DIRD are shown.

quires only 27 bytes (=216 bits / 8 bits per byte) of storage. In total we evaluate the continuous DIRD vectors which we refer to as DIRD (float), the byte-quantized version DIRD (byte) and the bit quantized binary version which we refer to as DIRD (bit).

We evaluate each of the DIRD flavors on three challenging test sets and compare its performance to state-of-the-art descriptors USURF and BRIEF. The USURF descriptor is run in extended mode which yields a 128 dimensional float vector. Moreover, we compare the DIRD family to a simple gray value descriptor which uses 256 gray values of pre-defined pixel positions around the interest point as the descriptor.

The first test set is the "day-to-night" test set taken from [30]. It shows the sky line of a city during the course of a full day. The test set consists of seven images in total where the first one is taken early in the morning and the last at night time. Three of these images are shown in Figure 6.11. The camera has been stationary over the full time span. From these test images we extract pairs of features corresponding to the same point but at different times. To this end, we first mask out the sky and thereafter compute an equidistantly spaced grid of pixel positions as can be seen in Figure 6.11. For each such grid point we extract a feature vector in every image. Pairs of corresponding feature vectors are found by computing all possible pairs showing the same spot. In total, 56000 positive feature pairs are computed thereby. Furthermore, we randomly pair features of different image location yielding yet another 56000 negative feature pairs. The performance of different descriptors is then assessed by computing the area under the ROC curve.

The results are depicted in Figure 6.12. The gray value descriptor that is used as a baseline is hardly able to distinguish positive from negative pairs as the illumination changes. Its AUC is 0.64. BRIEF performs significantly better and achieves an AUC of 0.89. USURF achieves a score of 0.92 showing some robustness to

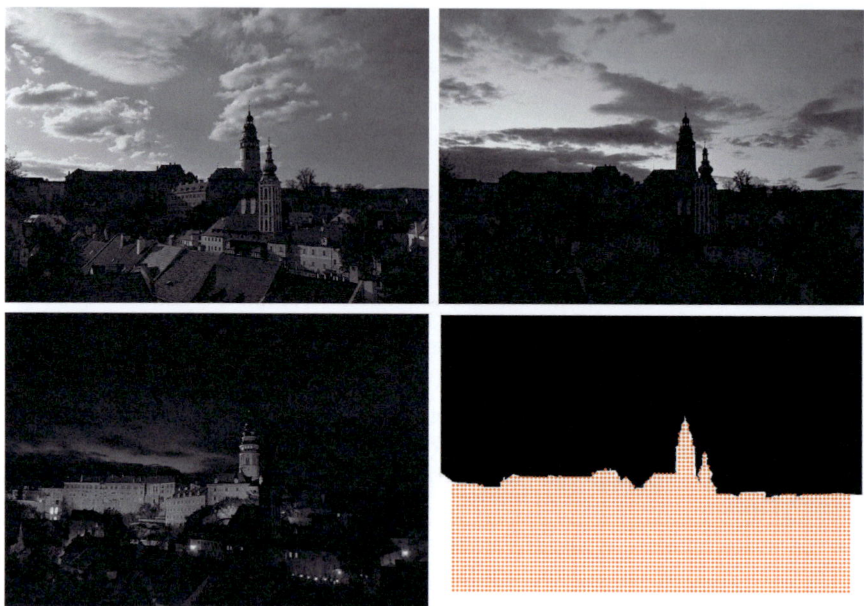

Figure 6.11: The "day-to-night" test set of [30] contains seven images taken over the course of a full day. Three sample images are shown. The sky is masked out and features are extracted for the marked (orange) pixel positions.

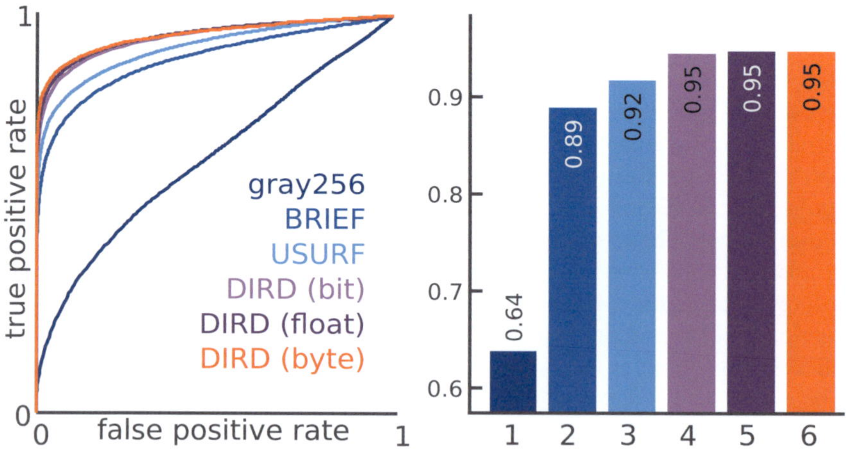

Figure 6.12: ROC curves for the tested descriptors are shown for the test set depicted in Figure 6.11. The bar plots show the respective AUCs.

this heavy illumination variation. The best scoring descriptors for this setup are, however, DIRD (bit), DIRD (float) and DIRD (byte) with values of 0.945, 0.948 and 0.949 respectively. Interestingly, all three perform almost equally well on this dataset despite a reduction in storage of factor 32 from DIRD (float) to DIRD (bit). We plan to further investigate the possibility of data reduction in the future (see Chapter 7).

Very recently, Zambanini and Kampel [71] have published work on illumination robust image descriptors and they kindly provided their dataset and we use it as a second test set. Image patches of a fixed size of 64×64 pixels are extracted from the Amsterdam library of object images (ALOI) [27]. These patches are grouped such that each group shows the same region of an object photographed under different illumination conditions. Some example patches are shown in Figure 6.13. We again, computed positive and negative image pairs and computed the corresponding feature vectors from them. In total we obtain 115000 patch pairs each. As before, descriptor performance is measured by the AUC.

Results for this dataset are shown in Figure 6.14. The result is comparable to the previous one. The gray value descriptor performs poorly with an AUC of 0.75 followed by BRIEF with an AUC of 0.88. USURF reaches an AUC of 0.93. DIRD (bit) and DIRD (float) perform equally well with an AUC of 0.95 as in the previous test. DIRD (byte), however, slightly outperforms them with an AUC of 0.96.

Finally, we assess the proposed descriptors by a real world point matching task. Two stereo images of different times showing the same place are taken for testing. The left images are shown in Figure 6.15. The top image is taken during morning hours whereas the second one was shot during evening hours. The local appearance is changed significantly between the two images. This, however, is a typical situation in which feature points need to be associated correctly for localization. In this test scenario we detect salient points in all four images (using both stereo images each). This detection of points is kept fixed for all subsequent experiments. Then points are associated by feature matching. One point is matched in a ring manner from image one to image four and then again to image one yielding a quadruple match. We keep only those matches for which the ring matching was successful i.e. the feature was correctly matched back to its initial pixel position in image one. Thereafter, we compute a visual odometry [26] estimate between these two stereo images and keep only those matches flagged as inliers by the visual odometry estimator. The matching results (inlier only) for the gray value descriptor, USURF, BRIEF, DIRD (bit), DIRD (float) and DIRD (byte) are shown in Figures 6.15 and 6.16 from top to bottom.

The baseline descriptor gray256 correctly matches only features that are very far away as their appearance does not change so much for this test. It matches 99 features. USURF performs second worse in this setup and correctly matches 550

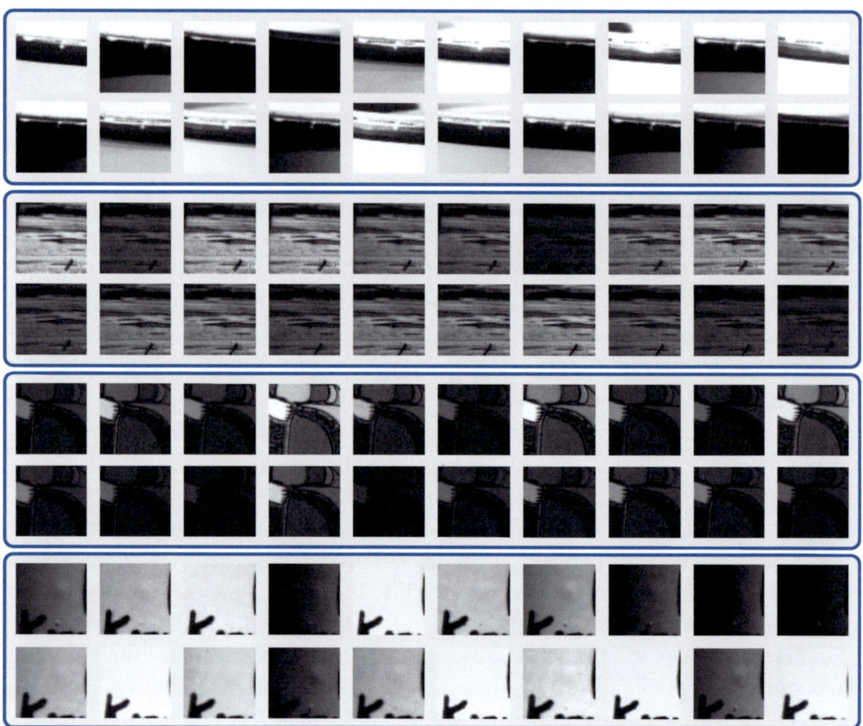

Figure 6.13: Some sample patches of the test set of [71] are shown. Corresponding patches are grouped. Positive and negative pairs are drawn and used in the evaluation.

interest points. Slightly more points (645) can be matched correctly using BRIEF. Our binary descriptor DIRD (bit) is able to correctly associate 992 points. Its continuous sibling DIRD (float) matches 1236 points and is only superseded by DIRD (byte) with 1391 inlier matches. These findings are summarized in Figure 6.17. The inlier ratio is also presented in Figure 6.17. It reveals that USURF and BRIEF match only 23% correctly whereas DIRD (byte) and DIRD (bit) match around 40% correctly. DIRD (float) exhibits an inlier ratio of 36%. The gray level descriptor is far off with 18%.
We note that the three DIRD variants are the top performing descriptors in all tests. Their increasing order is always DIRD (bit), DIRD (float) and DIRD (byte) except for the inlier fraction test where DIRD (bit) and DIRD (float) have changed places. We feel particularly encouraged to continue work on DIRD by the excellent point matching performance of DIRD (byte) which outperforms all alternatives by a

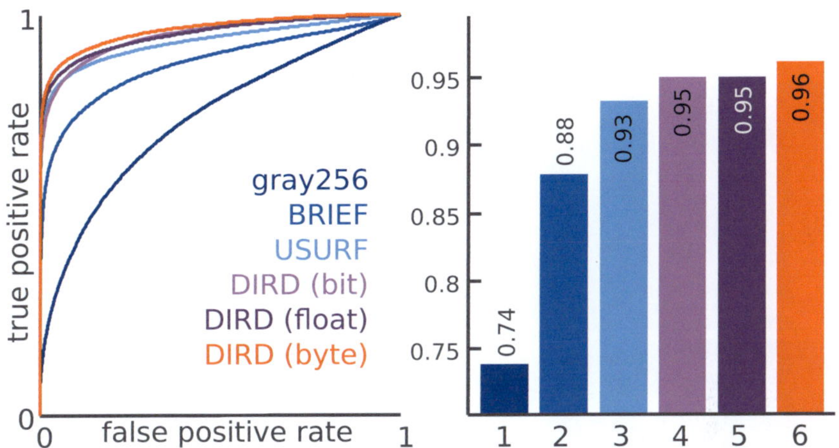

Figure 6.14: ROC curves and AUCs are shown for the tested descriptors for the test set shown in Figure 6.13. DIRD (byte) is the best performing descriptor.

substantial margin and more than doubles the performance of the best non-DIRD descriptor.

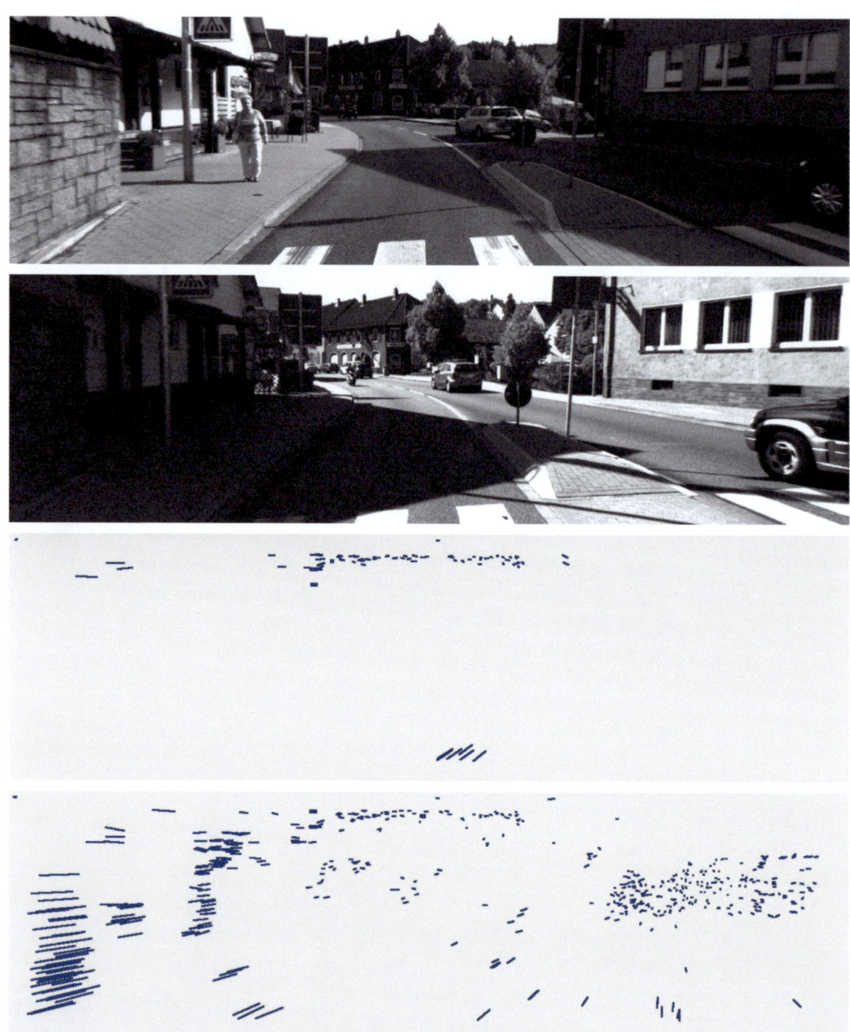

Figure 6.15: The top two images are used for a point matching task. The flow field of correctly associated points are shown for a simple gray value descriptor (99 inlier matches) and USURF (550 inlier matches).

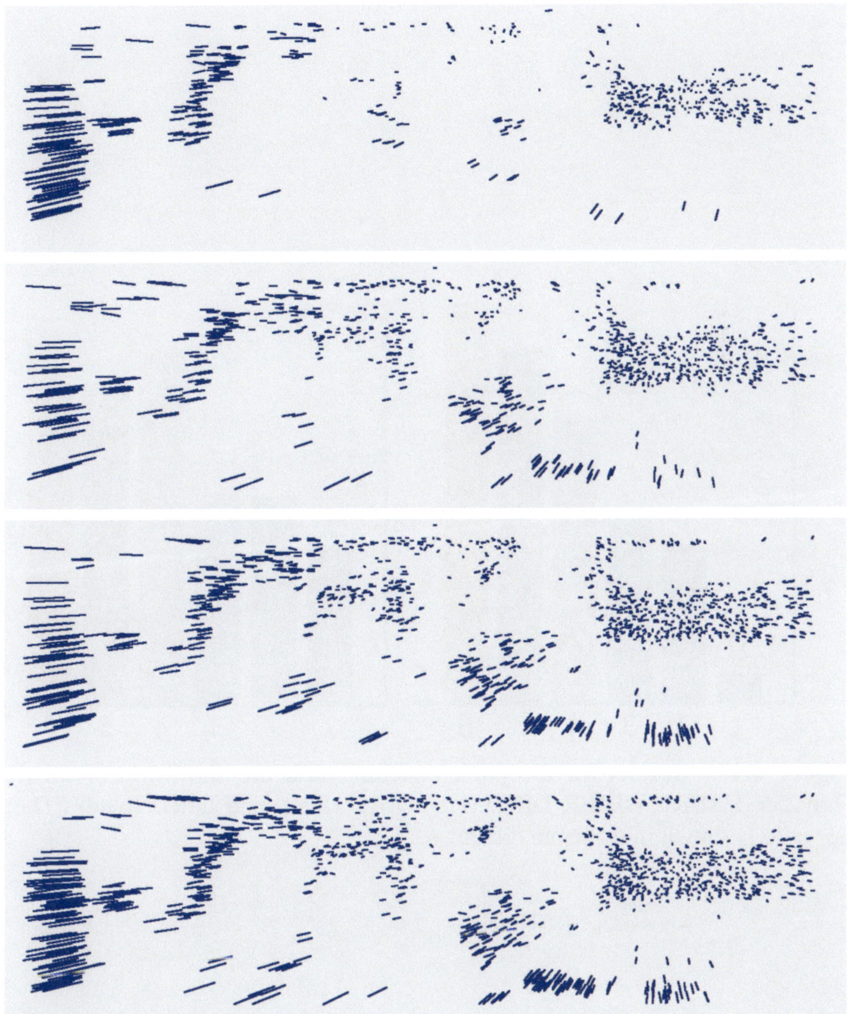

Figure 6.16: The flow fields of correctly associated points are shown for BRIEF (645 inlier matches), DIRD (bit) (922 inlier matches), DIRD (float) (1236 inlier matches) and DIRD (byte) (1392 inlier matches.)

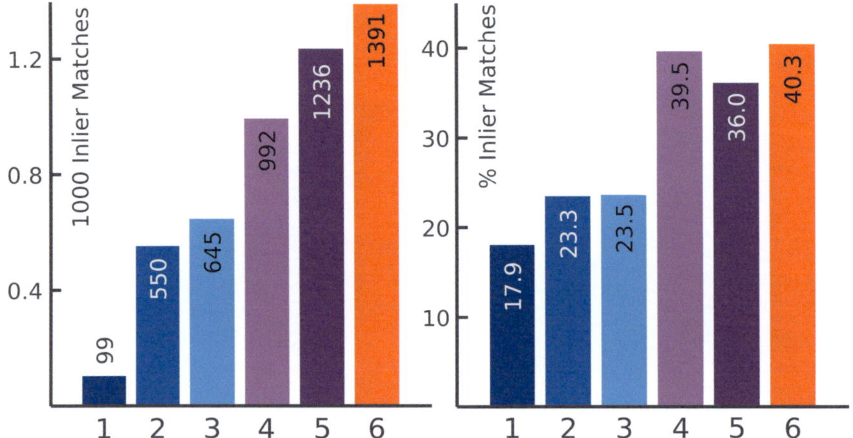

Figure 6.17: Results for the point matching task are shown. Left to right: Gray256, USURF, BRIEF, DIRD (bit), DIRD (flaot) and DIRD (byte). The inlier ratio is shown in the same order.

7 Conclusion

A set of algorithms for map relative visual localization using a single monocular camera was proposed in this thesis achieving centimeter level accuracy.

The mapping pipeline was presented in Chapter 4. It only requires a stereo camera and does not rely on any additional external hardware like odometers or GNSS receivers. Nevertheless, their optional inclusion is straight forward when desired. Stereo sequences are recorded such that they cover the area for which a map is sought. These sequences can be acquired on different days and times and the proposed algorithm merges these recordings and enforces consistency. Finally, landmark positions are computed and stored in an efficient map data structure.

The mapping process first computes a pose graph which consists of of all poses of all survey trajectories. Poses of the graph are interconnected by motion estimates computed from visual optometry. Motion estimates are computed for pairs of poses that are in close vicinity to each other. Consecutive poses of one survey trajectory are naturally considered to be neighbors. Self overlap and areas that overlap with other sequences are reliably and robustly found by the proposed place recognizer. It extracts a holistic feature vector for every single image of all mapping sequences. Thereafter, subsequences are matched to yield pairs of poses showing the same place. The subsequence matching algorithm is globally optimal and very efficient to compute. Once the pose graph is constructed it is solved by state-of-the-art NLS estimation to yield a pose representation that is consistent in all areas of self and cross overlap.

The place recognizer and pose graph optimizer was tested on a challenging data set of three survey trajectories with forward and backward facing cameras that contain numerous self loops each. Each sequence was tested in isolation first. Experiments show that all loops are correctly found by the proposed algorithm even in areas of visual ambiguity. Thereafter, a fourth image sequence was recorded that overlaps with each of the three earlier ones. All areas of overlap are robustly found and the jointly optimized pose graph consisting of all four recordings shows no false loop closure.

Next, a set of 3D landmarks is computed and added to the visual map. Salient image points are extracted from the recorded imagery and associated across all sequences. Hence, a single point in 3D is described by a set of pixel positions which are the projections of the point into their respective image planes. An NLS solver then extracts the 3D position of a given landmark that best explains these projections. A heuristic was proposed to prune some inappropriate landmarks from the map that will most likely not contribute to a reliable localization estimate.

Some experimental findings are presented showing bird eye views of the estimated landmark positions. Typical urban structures like parked cars can be seen from this representation. Back projecting the 3D landmarks into their image plane shows that an overwhelming majority of image points on moving objects have been identified as such and are not stored in the map data structure. Moreover, the mapping process scales well to large environments.

The proposed mapping pipeline was extensively tested on several hundred kilometers and the resulting maps were used successfully for localization. Nevertheless, a few improvements are possible to further increase both robustness and accuracy. First, a method to further reduce the storage requirements on the visual map appears useful as it is rather large. One solution may be to use only those landmarks that are successfully found in image sequences of different days as these appear to be persistent. Another exciting ground for future work is the development of a specialized key point detector that is specifically tailored to the road surface. This detector may even be computationally expensive as it is applied only during mapping. Experiments show that almost one magnitude more image points are detected during localization than there are landmarks in the map. This allows that almost all landmarks are matched. Thus, a specialized key point detector which is integrated into the mapping procedure could increase the localization robustness without any need to alter the localization algorithm.

The localization method that localizes a single monocular camera relative to the aforementioned visual map was presented in Chapter 5. It follows a three step approach. First, a topological localization is computed. The pose of the visual map that is closest to the current pose is found. This information is required to initialize the localizer. The proposed topological localization methods shows some resemblance to the place recognizer reviewed above. The current camera image is described holistically and the resulting feature vector is compared to all features of the visual map. Choosing the pose of the map that is the nearest pose in the space of appearances is susceptible to false alarms. Hence, a small set of promising candidate poses of the map are investigated by sequence matching which can be computed very efficiently. Thereby, the nearest pose of the map is found with great robustness.

The topological localization was tested on 500 equidistantly spaced starting positions of two localization test sets. The median distance until a topological localization is possible was determined. Despite a few rural areas where such methods require a substantially longer traveling distance the median is between seven and ten meters depending on the test set. The correctness of the topological localization was verified by a subsequent point feature matching and no single false topological localization was computed by our method.

The second step of the localization method computes a metric pose from point feature matches. All landmarks of the immediate vehicle vicinity (which is known from topological localization) and their associated visual descriptors are loaded from disk. Salient image points of the current camera image are associated with these 3D landmarks and NLS estimation and RANSAC are used to derive a six DOF ego pose estimate which we dubbed one-shot estimate.

Evaluating the cardinality of the inlier set of point matches for all images of all testing sequences shows that a reliable feature matching is possible despite naturally occurring changes of the environment like weather and illumination conditions. Moreover, we note that feature density and matching quantity varies and depends on the environment. The proposed method works best in urban environment.

Lastly, the localization algorithm stores a sliding window history of past one-shot estimates and re-optimizes these jointly. To this end, a motion model constraint is enforced and one-shot estimates serve as a prior in this optimization step. Hence, the prediction of a pose shall not deviate much from the pose estimate which in turn shall be close to its prior. As this set of constraints is overdetermined we again resort to NLS estimation to resolve it. This step of the localization algorithm is referred to as pose adjustment.

The visual map contains no global reference as it is constructed entirely from pose relative measurements. Hence, a localization accuracy evaluation cannot rely on any ground truth as it is unavailable. To assess the localization accuracy we propose to compare two independent localization estimates of two independent cameras mounted on the same vehicle for consistency. Both estimated trajectories are expected to deviate by a constant offset which precisely corresponds to the offset of the two cameras. Our experimental findings show a localization accuracy of a few centimeters and sub degree angular precision. Moreover, we presented qualitative experiments of an AR system which further supports these findings. Finally, we note that a slightly modified method has been used for automatic driving for several hundred kilometers in total.

The experiences we made with the proposed localization method are very encouraging. Nevertheless, we see some room to further improve this method. We expect the topological localization to benefit from a modified image description model. Using a bag-of-feature surrogate instead of the holistic approach will allow the method to work even when the camera mounting between mapping and localization is varied considerably. Following this line is especially tempting to try since point features of the camera image need to be computed for metric localization already and are hence obtained without any additional computational expenses. Moreover, the loss of realtime capability of the topological localization is inevitable once the map size becomes vast (e.g. city scale and beyond). However, state-of-the-art image retrieval methods [33] offer a solution. A key technology seems to be a combination of hashing and efficiently storing feature vectors by means of binary quantization. We believe that the investigation of this branch of algorithms for topological localization is a fruitful direction of future work. Lastly, we cannot resist to extend the pose adjustment to use more advanced motion models such the curve linear models of [60]. These, however, require the coordinate transform from camera to vehicle coordinates to be available. We plan to estimate this transform jointly with the poses during pose adjustment which would extend the associated factor graph into a hyper graph representation.

Descriptors are widely used in many computer vision problems and a construction scheme to automatically assemble them was presented in Chapter 6. A set of algorithmic building blocks are introduced which can be chained in any order. Each such combination of building blocks (processing steps) yields one descriptor each. It was shown how commonly used descriptors like BRIEF and LBP can be computed using these blocks. The construction of HOG and SURF is elucidated in Appendix A. To evaluate any given descriptor we have introduced a fitness function to assess the suitability for place recognition under varying illumination conditions. Then a meta heuristic was used to evaluate an initial population of random descriptors. The best performing ones are slightly modified by mutation and form the next generation. Evaluation, selection and mutation are cycled until convergence. Thereby we were able to fully automatically learn a novel descriptor which is specifically designed to be robust to illumination variations and we dubbed it DIRD. The performance of DIRD was compared to state-of-the-art image descriptors like SURF, U-SURF, BRIEF and a simple gray value descriptor. DIRD outperformed all handcrafted alternatives on an independent test set by a significant margin. This finding speaks much in the favor of using specialized image descriptors over general purpose ones. Moreover, DIRD can be computed efficiently and is used for both place recognition and landmark feature matching in all experiments.

An obvious next step is to test the feature learning method on other computer vision problems. Preliminary tests on point matching benchmarks seem promising and we plan to further pursue this idea. An important issue that is neglected by the proposed feature algorithm learning procedure is the unbound complexity of the resulting descriptor. The automatically learned DIRD is light weight even though we believe this has merely been a lucky chance. Nevertheless, enforcing computational efficiency does not appear to be a far cry from what is presented in Chapter 6. Each building block increases the computational complexity of the descriptor in a deterministic and predictable manner. Hence, descriptors of high computational cost can be evaluated artificially poor by the fitness function forcing the learning framework to favor light weight chains of building blocks over heavy weight ones.

A   Appendix

A set of building blocks with block specific variants are presented in Chapter 6 which can be chained in an arbitrary order such that one such combination of blocks yields one descriptor each. The set of blocks allows to span a large set of different descriptors. BRIEF and LBP are nowadays commonly used ones and it was shown in Chapter 6 how these can be assembled by the proposed set of blocks. Herein, we show how the HOG [18] descriptor which is the work horse of pedestrian detection and the SURF [6] descriptor which is the most commonly used general purpose feature extractor can be constructed by the proposed blocks. HOG is constructed in Section A.1 and the construction of SURF is presented in Section A.2.

A.1   Histograms of Oriented Gradients

HOG features are widely used in pedestrian detection and classification. Next, we show how HOG features are constructed by using the algorithmic building blocks of Chapter 6. To this end, we first briefly review the method as presented in [18]. The region for which a HOG feature shall be computed is first partitioned into small squared patches of size 4×4 pixels. Such a small patch is referred to as a cell. For each pixel of a cell the derivative in horizontal and vertical direction are computed and denoted by dx and dy. Thereafter, the derivatives are transformed into a polar representation yielding orientation $\theta = \mathrm{atan2}(dy, dx)$ and magnitude $m = \sqrt{dx^2 + dy^2}$ for every pixel of the cell. A gradient histogram is thereafter computed for the particular cell which has six bins. Note, that the original method proposes to weight the orientation by the magnitude to compute the histogram. Herein, however, we chose to compute a non-weighted histogram using the orientation information only. Next, cells are merged into blocks each of which consist of 2×2 cells each. The orientation histograms of the cells of one block are concatenated to form one feature vector for each block which is thereafter scaled to unit norm. Several such block feature vectors are computed by shifting the block by a fixed pixel count (referred to as block stride) and re-evaluating the block feature. This shifting procedure is continued until the entire patch is covered. All block features are finally concatenated to form the HOG feature. We finally note that it is possible to vary many parameters such as block stride, number of histogram bins, cell and block sizes etc. For simplicity we have presented a default set of parameters here.

Next, we present how to assemble the HOG descriptor from our set of building blocks. To this end, we describe the chain of blocks and explain their respective variants. First, we apply the derivative filter bank to yield a 2-vector $(dx, dy)^T$ for the pixel position the descriptor is applied to. The nonlinear transform that translates Cartesian to polar coordinates is readily available and applied thereafter to yield $(m, \theta)^T$. The histogram of orientations is computed from the histogram block. It uses the neighborhood $\mathcal{O}_N$ that covers the area of the cell and has histogram centers $\mathcal{C} = (c_1, \ldots, c_6)$ with $c_i = (0, i\pi/3)^T$ where the magnitude is always zero and the orientations equidistantly cover one full rotation. Thereby, the magnitude is irrelevant and an orientation only histogram is computed. The repetition to block level which concatenates several such histograms for a block is achieved by the repetition building block. The block feature is scaled to unit norm by the normalization building block. The final HOG feature is computed by applying the thus constructed descriptor at several positions covering the entire patch which corresponds to the repetition operation.

HOG can be expressed by summarizing the construction steps with our building block notation:

1. Apply derivative filter bank.

2. Nonlinearly transform from Cartesian to Polar coordinates.

3. Compute a histogram over the area of a cell with bin centers at equidistantly spaced angles covering one full rotation.

4. Apply repetition to yield block level features.

5. Apply normalization building block to scale feature to unit norm.

6. Apply yet another repetition pattern to cover the entire patch.

Finally, we note that the original magnitude weighted histograms cannot be constructed by the building blocks of Chapter 6 without modification.

A.2   Speeded Up Robust Feature

It is largely agreed that SURF is the most successful feature descriptor to date as it is fairly quick to compute and shows good general purpose performance. We describe how SURF can be assembled by our building blocks.

We assume that a pixel position u, v of an image I has been detected by some detector. Moreover, we assume that the scale and orientation of the salient point has been correctly detected by the detector. This is the common case where rotation and scale invariance is required. These invariance proprieties are attributed to the detector and not the descriptor. Henceforth, we assume a square patch of size $s \times s$ pixels which is axis aligned and thus compensated for scale and orientation and we show how the patch around the center pixel u, v can be described by SURF. To this end, we briefly restate the method presented in [6] before translating it into our building blocks vocabulary.

The square path of size $s \times s$ is tiled into 4×4 equally sized, non-overlapping sub-patches. For each sub-patch a set of 5×5 equally spaced pixel positions are determined such that they form a grid covering the area of the sub-patch. A two dimensional Haar feature $(dx_{p,k,l}, dy_{p,k,l})$ is computed for each pixel position $(k, l) \in \{1, \dots, 5\} \times \{1, \dots, 5\}$ of the grid where $p \in \{1, \dots, 16\}$ denotes the sub-patch id. These Haar features are horizontal and vertical Haar features of a fixed scale. Thereafter, each Haar feature is weighted by a weight $w_{p,k,l}$ which depends on the distance of the pixel to the center pixel u, v of the patch. Thereafter, a four-vector is computed for each sub-patch p by

$$
v_p = \begin{pmatrix} \sum_{k,l} w_{p,k,l} dx_{p,k,l} \\ \sum_{k,l} w_{p,k,l} dy_{p,k,l} \\ \sum_{k,l} |w_{p,k,l} dx_{p,k,l}| \\ \sum_{k,l} |w_{p,k,l} dy_{p,k,l}|) \end{pmatrix} \tag{A.1}
$$

where k, l sums over all twenty five grid pixels of the sub-patch p. Last all sub-patch vectors v_p are concatenated and the final vector is scaled to unit norm yielding the 64 dimensional SURF feature.

Next we rephrase the aforementioned method using our blocks and variants notation. The first set of blocks will perform on only one grid pixel position of one sub-patch. The processing pipeline starts with the Haar feature filter bank. It will compute Haar features for several scales $n = 1, \dots, N$ in both vertical, horizontal and diagonal direction. So far, the descriptor will compute a feature $f = (dx_1, dy_1, dxy_1, \dots, dx_N, dy_N, dxy_N)^T$ for each pixel it is applied to where dx_n and dy_n are horizontal and vertical Haar features as above and dxy_n is the diagonal Haar feature of scale n. For SURF, however, we are interested in only one particular scale n' (which is a fixed parameter) and only the vertical and horizontal features. To this end we subsequently apply a linear transform by multiplying the feature vector with a matrix A. The matrix $A \in \mathbb{R}^{2 \times 3N}$ contains exactly one 1 in each row and zeros otherwise. Thereby the feature obtained by matrix multiplication yields $Af = (dx_{n'}, dy_{n'})^T$ as required. One can think of A as a "selection matrix" which picks those components of the feature vector that are required by SURF.

Next we apply the summation block to each pixel the descriptor Af is applied to. The neighborhood $\mathcal{O}_N$, however, consists only of a single pixel position offset which is zero. Thereby the summation yields a new feature which is

$$\begin{pmatrix} \sum_{\mathcal{N}} dx_{n'} \\ \sum_{\mathcal{N}} dy_{n'} \\ \sum_{\mathcal{N}} |dx_{n'}| \\ \sum_{\mathcal{N}} |dy_{n'}| \end{pmatrix} = \begin{pmatrix} dx_{n'} \\ dy_{n'} \\ |dx_{n'}| \\ |dy_{n'}| \end{pmatrix} \tag{A.2}$$

because $\mathcal{O}_N$ is only one pixel in size. The descriptor constructed so far will output the four vector of (A.2) for a given pixel position.

This descriptor is now extended as follows. A repetition is the next processing block. Thus, a new descriptor is constructed by using the thus far assembled one by applying it on all 20×20 grid pixel positions of all sub patches and concatenating the outputs. This will yield a (temporary) large feature vector of dimension $20 \cdot 20 \cdot 4$. Next, the weighting with the fixed weights $w_{j,k,p}$ is integrated into our chain of building blocks. To this end, the feature vector is multiplied by a large diagonal matrix $B = \mathrm{diag}(w_{1,1,1}, w_{1,1,1}, w_{1,1,1}, w_{1,1,1}, \ldots, w_{5,5,16})$ which contains the weights. Note that each weight is repeated four times to weight each component of the concatenated four vectors of (A.2).

Finally, the sub patch summation of SURF is achieved by another linear transform. The weighted feature is multiplied by yet another matrix C which contains 25 ones per row so that the features of one sub patch are summed each. The final step of unit length normalization is readily contained in our set of building blocks and hence applied last.

SURF can be expressed by summarizing the construction steps with our building block notation:

1. Apply Haar filter bank.

2. Linear Transform with matrix A to select the two appropriate Haar features.

3. Summation over single pixel neighborhood to yield the vector of (A.2).

4. Repeat the thus constructed descriptor on all 20×20 grid pixel positions of all sub-patches and concatenate the output.

5. Use diagonal matrix B in another linear transform to integrate the weights for each feature.

6. Use yet another linear transform with matrix C to achieve the sub-patch summation.

7. Apply normalization block.

Obviously, the two subsequent linear transforms could be summarized by one transform with matrix CB but is kept separate here for better readability. Note, that all matrices are fixed and are completely independent of the pixel position u, v.

Bibliography

[1] P. Agarwal and E. Olson. Variable reordering strategies for slam. In *IEEE/RSJ International Conference on Intelligent Robots and Systems (IROS)*, 2012.

[2] P. Alcantarilla, O. Stasse, S. Druon, L. Bergasa, and F. Dellaert. How to localize humanoids with a single camera? *Autonomous Robots*, 2012.

[3] H. Badino, D. Huber, and T. Kanade. Visual topometric localization. In *IEEE Intelligent Vehicles Symposium (IV)*, pages 794–799, 2011.

[4] H. Badino, D. Huber, and T. Kanade. Real-time topometric localization. In *IEEE International Conference on Robotics and Automation (ICRA)*, pages 1635–1642, 2012.

[5] T. Bailey and H. Durrant-Whyte. Simultaneous localization and mapping (slam): Part ii. *IEEE Robotics & Automation Magazine*, 13(3):108–117, 2006.

[6] H. Bay, T. Tuytelaars, and L. Van Gool. Surf: Speeded up robust features. In *European Conference on Computer Vision (ECCV)*, pages 404–417, 2006.

[7] J. Beck. Automatisiertes Deskriptorlernen zur Ortserkennung mit Kameras *Diplomarbeit KIT*, 2012.

[8] D. Bertsekas. Incremental least squares methods and the extended kalman filter. *SIAM Journal on Optimization*, 6(3):807–822, 1996.

[9] H. Bcyer. *The theory of evolution strategies*. Spring, 2001.

[10] E. Bingham and H. Mannila. Random projection in dimensionality reduction: applications to image and text data. In *International conference on Knowledge discovery and data mining*, pages 245–250, 2001.

[11] Y. Boureau, F. Bach, Y. LeCun, and J. Ponce. Learning mid-level features for recognition. In *IEEE Conference on Computer Vision and Pattern Recognition (CVPR)*, pages 2559–2566, 2010.

[12] M. Brown, G. Hua, and S. Winder. Discriminative learning of local image descriptors. *IEEE Transactions on Pattern Analysis and Machine Intelligence (PAMI)*, 33(1):43–57, 2011.

[13] M. Calonder, V. Lepetit, C. Strecha, and P. Fua. Brief: binary robust independent elementary features. In *European Conference on Computer Vision (ECCV)*, pages 778–792, 2010.

[14] G. Carneiro. The automatic design of feature spaces for local image descriptors using an ensemble of non-linear feature extractors. In *IEEE Conference on Computer Vision and Pattern Recognition (CVPR)*, pages 3509–3516, 2010.

[15] O. Chum and J. Matas. Optimal randomized ransac. *IEEE Transactions on Pattern Analysis and Machine Intelligence (PAMI)*, 30(8):1472–1482, 2008.

[16] Eurostat European Commision. Europe in figures. 2012.

[17] M. Cummins and P. Newman. Fab-map: Probabilistic localization and mapping in the space of appearance. *International Journal of Robotics Research*, 27(6):647–665, 2008.

[18] N. Dalal and B. Triggs. Histograms of oriented gradients for human detection. In *IEEE Conference on Computer Vision and Pattern Recognition (CVPR)*, pages 886–893, 2005.

[19] A. Davison, I. Reid, N. Molton, and O. Stasse. Monoslam: Real-time single camera slam. *IEEE Transactions on Pattern Analysis and Machine Intelligence (PAMI)*, 29(6):1052–1067, 2007.

[20] F. Dellaert, J. Carlson, V. Ila, K. Ni, and C. Thorpe. Subgraph-preconditioned conjugate gradients for large scale slam. In *IEEE/RSJ International Conference on Intelligent Robots and Systems (IROS)*, pages 2566–2571, 2010.

[21] V. Drevelle and P. Bonnifait. Global positioning in urban areas with 3-d maps. In *IEEE Intelligent Vehicles Symposium (IV)*, pages 764–769, 2011.

[22] Statistisches Bundesamt DSTATIS. Transport und verkehr. *Statistisches Jahrbuch 2012*, 2012.

[23] H. Durrant-Whyte and T. Bailey. Simultaneous localization and mapping: part i. *IEEE Robotics & Automation Magazine*, 13(2):99–110, 2006.

[24] M. Fischler and R. Bolles. Random sample consensus: a paradigm for model fitting with applications to image analysis and automated cartography. *Communications of the ACM*, 24(6):381–395, 1981.

[25] S. Gauglitz, T. Höllerer, and M. Turk. Evaluation of interest point detectors and feature descriptors for visual tracking. *International journal of computer vision*, pages 1–26, 2011.

[26] A. Geiger, J. Ziegler, and C. Stiller. Stereoscan: Dense 3d reconstruction in real-time. In *IEEE Intelligent Vehicles Symposium (IV)*, pages 963–968, 2011.

[27] J. Geusebroek, G. Burghouts, and A. Smeulders. The amsterdam library of object images. *International Journal of Computer Vision (IJCV)*, 61(1):103–112, 2005.

[28] G. Grisetti, R. Kümmerle, C. Stachniss, and W. Burgard. A tutorial on graph-based slam. *IEEE Intelligent Transportation Systems Magazine*, 2(4):31–43, 2010.

[29] R. Hartley and A. Zisserman. *Multiple view geometry in Computer Vision*. Cambridge University Press, 2000.

[30] J. Heinly, E. Dunn, and J. Frahm. Comparative evaluation of binary features. In *European Conference on Computer Vision (ECCV)*, pages 759–773, 2012.

[31] C. Hertzberg. A framework for sparse, non-linear least squares problems on manifolds *diplomarbeit universiät bremen* , 2008.

[32] B. Jähne. *Digitale Bildverarbeitung*. Springe, 2005.

[33] H. Jegou, F. Perronnin, M. Douze, C. Schmid, et al. Aggregating local image descriptors into compact codes. *IEEE Transactions on Pattern Analysis and Machine Intelligence (PAMI)*, 34(9):1704–1716, 2012.

[34] F.R. Kschischang, B.J. Frey, and H.A. Loeliger. Factor graphs and the sum-product algorithm. *IEEE Transactions on Information Theory*, 47(2):498–519, 2001.

[35] R. Kuemmerle, G. Grisetti, H. Strasdat, K. Konolige, and W. Burgard. g2o: A general framework for graph optimization. In *IEEE International Conference on Robotics and Automation (ICRA)*, pages 3607–3613, 2011.

[36] H. Lategahn, J. Beck, B. Kitt, and C. Stiller. How to learn an illumination robust image feature for place recognition. In *IEEE Intelligent Vehicles Symposium (IV)*, Gold Coast, Australia, 2013.

[37] H. Lategahn, A. Geiger, and B. Kitt. Visual slam for autonomous ground vehicles. In *IEEE International Conference on Robotics and Automation (ICRA)*, Shanghai, China, 2011.

[38] H. Lategahn, A. Geiger, B. Kitt, and C. Stiller. Motion-without-structure: Real-time multipose optimization for accurate visual odometry. In *IEEE Intelligent Vehicles Symposium (IV)*, Alcala de Henares, Spain, 2012.

[39] H. Lategahn, T. Graf, C. Hasberg, B. Kitt, and J. Effertz. Mapping in dynamic environments using stereo vision. In *IEEE Intelligent Vehicles Symposium (IV)*, Baden-Baden, Germany, 2011.

[40] H. Lategahn, S. Gross, T. Stehle, and T. Aach. Texture classification by modeling joint distributions of local patterns with gaussian mixtures. *IEEE Transactions Image Processing*, 19(6):1548–1557, 2010.

[41] H. Lategahn, M. Schreiber, J. Ziegler, and C. Stiller. Urban localization with camera and inertial measurement unit. In *IEEE Intelligent Vehicles Symposium (IV)*, Gold Coast, Australia, 2013.

[42] H. Lategahn and C. Stiller. City gps using stereo vision. In *IEEE International Conference on Vehicular Electronics and Safety (ICVES)*, Istanbul, Turkey, 2012.

[43] H. Lategahn and C. Stiller. Experimente zur hochpräzisen landmarken-basierten eigenlokalisierung in unsicherheitsbehafteten digitalen karten. In *FAS Workshop*, 2012.

[44] H. Lategahn, A. Wege, T. Graf, J. Effertz, and B. Kitt. Schnelle Berechnung von detaillierten Belegungsgittern aus dichten Stereodisparitätsbildern. In *Sicherheit durch Fahrerassistenz, TÜV Süd*, 2010.

[45] J. Levinson, M. Montemerlo, and S. Thrun. Map-based precision vehicle localization in urban environments. In *Robotics: Science and Systems Conference (RSS)*, 2007.

[46] J. Levinson and S. Thrun. Robust vehicle localization in urban environments using probabilistic maps. In *IEEE International Conference on Robotics and Automation (ICRA)*, pages 4372–4378, 2010.

[47] H. Li, F. Nashashibi, and G. Toulminet. Localization for intelligent vehicle by fusing mono-camera, low-cost gps and map data. In *IEEE International Conference on Intelligent Transportation Systems (ITSC)*, pages 1657–1662, 2010.

[48] D. Lowe. Distinctive image features from scale-invariant keypoints. *International journal of computer vision*, 60(2):91–110, 2004.

[49] F. Lu and E. Milios. Globally consistent range scan alignment for environment mapping. *Autonomous robots*, 4(4):333–349, 1997.

[50] M. Milford. Visual route recognition with a handful of bits. In *Robotics: Science and Systems Conference (RSS)*, 2012.

[51] A. Napier and P. Newman. Generation and exploitation of local orthographic imagery for road vehicle localisation. In *IEEE Intelligent Vehicles Symposium (IV)*, pages 590–596, 2012.

[52] A. Napier, G. Sibley, and P. Newman. Real-time bounded-error pose estimation for road vehicles using vision. In *IEEE International Conference on Intelligent Transportation Systems (ITSC)*, pages 1141–1146, 2010.

[53] T. Ojala, M. Pietikainen, and T. Maenpaa. Multiresolution gray-scale and rotation invariant texture classification with local binary patterns. *IEEE Transactions on Pattern Analysis and Machine Intelligence (PAMI)*, 24(7):971–987, 2002.

[54] J. Philbin, M. Isard, J. Sivic, and A. Zisserman. Descriptor learning for efficient retrieval. *European Conference on Computer Vision (ECCV)*, pages 677–691, 2010.

[55] P. Piniés and J. Tardós. Large-scale slam building conditionally independent local maps: Application to monocular vision. *IEEE Transactions on Robotics*, 24(5):1094–1106, 2008.

[56] O. Pink. Visual map matching and localization using a global feature map. In *IEEE Computer Vision and Pattern Recognition Workshops*, 2008.

[57] M. Powell. An efficient method for finding the minimum of a function of several variables without calculating derivatives. *The computer journal*, 7(2):155–162, 1964.

[58] W. Press, S. Teukolsky, W. T Vetterling, and B. Flannery. Numerical recipes in c: the art of scientific computing. 2. *Cambridge: CUP*, 1992.

[59] M. Schreiber, C. Knoeppel, and U. Franke. Laneloc: Lane marking based localization using highly accurate maps. In *IEEE Intelligent Vehicles Symposium (IV)*, Gold Coast, Australia, 2013.

[60] R. Schubert, E. Richter, and G. Wanielik. Comparison and evaluation of advanced motion models for vehicle tracking. In *IEEE International Conference on Information Fusion*, 2008.

[61] G. Sibley, L. Matthies, and G. Sukhatme. A sliding window filter for incremental slam. *Unifying perspectives in computational and robot vision*, pages 103–112, 2008.

[62] G. Sibley, C. Mei, I. Reid, and P. Newman. Vast-scale Outdoor Navigation Using Adaptive Relative Bundle Adjustment. *International Journal of Robotics Research*, 2010.

[63] R. Smith, M. Self, and P. Cheeseman. Estimating uncertain spatial relationships in robotics. *Autonomous robot vehicles*, 1:167–193, 1990.

[64] D. Stavens and S. Thrun. Unsupervised learning of invariant features using video. In *IEEE Conference on Computer Vision and Pattern Recognition (CVPR)*, pages 1649–1656, 2010.

[65] N. Sunderhauf, M. Obst, G. Wanielik, and P. Protzel. Multipath mitigation in gnss-based localization using robust optimization. In *IEEE Intelligent Vehicles Symposium (IV)*, pages 784–789, 2012.

[66] N. Sunderhauf and P. Protzel. Brief-gist-closing the loop by simple means. In *IEEE/RSJ International Conference on Intelligent Robots and Systems (IROS)*, pages 1234–1241, 2011.

[67] S. Thompson, S. Kagami, and M. Okajima. Constrained 6dof localisation for autonomous vehicles. In *IEEE International Conference on Systems Man and Cybernetics (SMC)*, pages 330–335, 2010.

[68] B. Triggs, P. McLauchlan, R. Hartley, and A. Fitzgibbon. Bundle adjustment - a modern synthesis. *Vision algorithms: theory and practice*, pages 153–177, 2000.

[69] A. Wendel, A. Irschara, and H. Bischof. Natural landmark-based monocular localization for mavs. In *IEEE International Conference on Robotics and Automation (ICRA)*, pages 5792–5799, 2011.

[70] S. Winder, G. Hua, and M. Brown. Picking the best daisy. In *IEEE Conference on Computer Vision and Pattern Recognition (CVPR)*, pages 178–185, 2009.

[71] S. Zambanini and M. Kampel. A local image descriptor robust to illumination changes. In *Conference on Image Analysis*, 2013.